Math Connects

Reteach and Skills Practice Workbook

Course 1

Also available online at

connectED.mcgraw-hill.com

To the Teacher These worksheets are the same as those found in the Chapter Resource Masters for Glencoe's *Math Connects*, Course 1. The answers to these worksheets are available at the end of each Chapter Resource Masters booklet.

The McGraw·Hill Companies

 Glencoe

Send all inquiries to:
Glencoe/McGraw-Hill
8787 Orion Place
Columbus, OH 43240-4027

ISBN: 978-0-07-895133-6
MHID: 0-07-895133-X

Reteach and Skills Practice Workbook, Course 1

Printed in the United States of America.

9 10 11 12 13 QVS 19 18 17 16 15

CONTENTS

Chapter 1 Multiply and Divide Decimals

Chapter 2 Multiply and Divide Fractions

Chapter 3 Ratios and Rates

Chapter 4 Fractions, Decimals, and Percents

Chapter 5 Algebraic Expressions

Chapter 6 Equations

Chapter 7 Functions, Inequalities, and Integers

Chapter 10 Volume and Surface Area

Chapter 11 Analyze Data and Graphs

Chapter 12 Probability

Multi-Part Lesson 1 PART A

NAME ______________________ DATE ____________ PERIOD ______

Reteach

Estimate Products

When estimating a product, round to the nearest whole number by looking at the digit in the tenths place and rounding down when the digit is 4 or less and up when the digit is 5 or higher.

Example 1 **Estimate a Product**

5.8 × 7

$$\begin{array}{r} 5.8 \\ \underline{\times 7} \end{array} \rightarrow \begin{array}{r} 6 \\ \underline{\times 7} \\ 42 \end{array}$$

Round 5.8 to 6.

The product is *about* 42.

Example 2 **Estimate a Product**

Estimate. 9.6 × 4.2

$$\begin{array}{r} 9.6 \\ \underline{\times 4.2} \end{array} \rightarrow \begin{array}{r} 10 \\ \underline{\times 4} \\ 40 \end{array}$$

Round 9.6 to 10.
Round 4.2 to 4.

The product is *about* 40.

Exercises

Estimate each product.

1. 7.4 × 2.7

2. 9.2 × 8.8

3. 22.1 × 9.9

4. 1.1 × 7.5

5. 14.4 × 8

6. 37.2 × 7

7. 19.6 × 5.4

8. 64.3 × 3.8

9. $\begin{array}{r} 35.1 \\ \underline{\times 10.2} \end{array}$

10. $\begin{array}{r} 12.6 \\ \underline{\times 3.2} \end{array}$

NAME ______________________ DATE ____________ PERIOD _____

Skills Practice

Estimate Products

Estimate each product.

1. 2.8×7.9

2. 72.1×49.7

3. 21.2×14

4. 3.8×11

5. 53.4×8.1

6. 15.7×6.2

7. 36.3×4

8. 88.5×3

9. 43.9×5.6

10. 18.4×2.7

11. 25.8×6.5

12. 61.2×7

Use estimation to determine whether each answer is reasonable. If the answer is reasonable, write *yes*. If not, write *no* and provide a reasonable estimate.

13. $110.9 \times 71 = 7{,}700$

14. $42.9 \times 101 = 4{,}300$

Multi-Part Lesson 1 PART C

NAME ______________________ DATE ____________ PERIOD ______

Reteach

Multiply Decimals by Whole Numbers

When you multiply a decimal by a whole number, you multiply the numbers as if you were multiplying all whole numbers. Then you use estimation or you count the number of decimal places to decide where to place the decimal point. If there are not enough decimal places in the product, annex zeros to the left.

Example 1 **Find 5 × 0.36 using models.**

Model 5 groups of 0.36 on a decimal model.

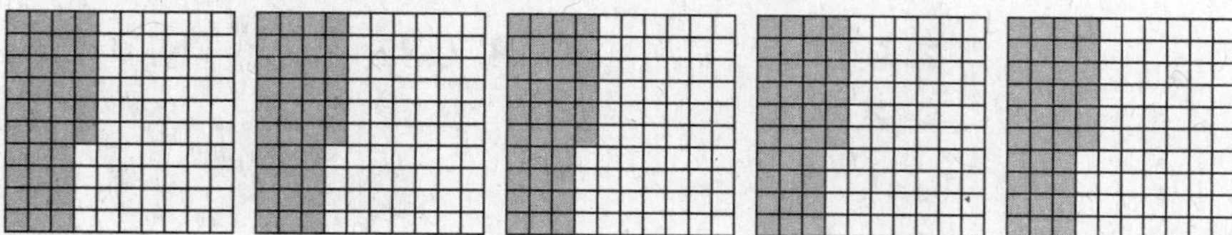

Rearrange the columns and small squares to fill as many whole grids as possible.

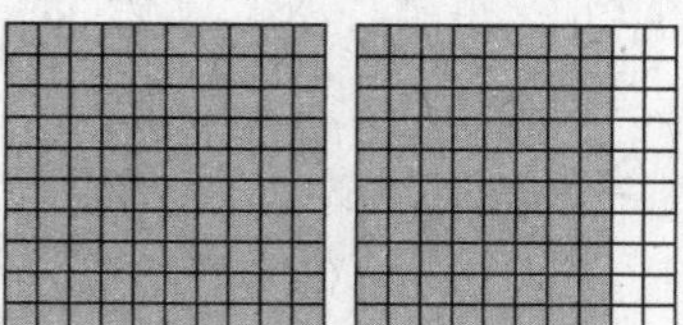

So, 5 × 0.36 = 1.8.

Example 2 **Find 6.25 × 5.**

Method 1 Use estimation.

Round 6.25 to 6.

6.25 × 5 → 6 × 5 or 30

```
   1 2
  6.25
×    5
 31.25
```

Since the estimate is 30, place the decimal point after 31.

Method 2 Count decimal places.

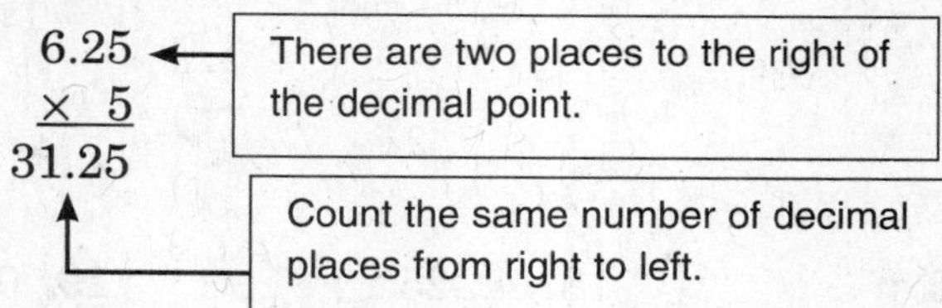

```
 6.25  ← There are two places to the right of the decimal point.
×   5
31.25
↑ Count the same number of decimal places from right to left.
```

Example 3 **Find 3 × 0.0047.**

```
     2
0.0047
×    3
0.0141
```

There are four decimal places.

Annex a zero on the left of 141 to make four decimal places.

Exercises

Multiply.

1. 8.03 × 3 **2.** 6 × 12.6 **3.** 2 × 0.012 **4.** 0.0008 × 9

5. 2.32 × 5 **6.** 6.8 × 7 **7.** 5.2 × 4 **8.** 1.412 × 3

Multi-Part Lesson 1

PART C

NAME ______________ DATE __________ PERIOD _____

Skills Practice

Multiply Decimals by Whole Numbers

Multiply. Use models if needed.

1. 1.5 × 3

2. 0.9 × 6

3. 0.45 × 5

4. 3.12 × 8

5. 3.47 × 5

6. 2.08 × 6

7. 9.14 × 2

8. 0.82 × 9

9. 6.3 × 9

10. 0.02 × 3

11. 9.12 × 4

12. 27.3 × 8

13. 4.007 × 4

14. 3.13 × 3

15. 5.02 × 8

16. 6.31 × 6

17. 8.01 × 5

18. 4.325 × 7

19. 0.762 × 2

20. 0.08 × 8

21. 6 × 3.04

22. 2.6 × 9

23. 13 × 2.5

24. 1.006 × 4

Multi-Part Lesson 1 PART E

NAME ______________________ DATE ____________ PERIOD _____

Reteach

Multiply Decimals by Decimals

When you multiply a decimal by a decimal, multiply the numbers as if you were multiplying all whole numbers. To decide where to place the decimal point, find the sum of the number of decimal places in each factor. The product has the same number of decimal places.

Example 1 **Find 5.2 × 6.13.**

Estimate: 5 × 6 or 30

```
    5.2  ←— one decimal place
 × 6.13  ←— two decimal places
    156
    52
 +312
 31.876  ←— three decimal places
```

The product is 31.876. Compared to the estimate, the product is reasonable.

Example 2 **Find 2.3 × 0.02.**

Estimate: 2 × 0.02 or 0.04

```
    2.3  ←— one decimal places
 × 0.02  ←— two decimal place
  0.046  ←— Annex a zero to make three decimal places.
```

The product is 0.046. Compared to the estimate, the product is reasonable.

Exercises

Multiply.

1. 7.2 × 2.1

2. 4.3 × 8.5

3. 2.64 × 1.4

4. 14.23 × 8.21

5. 5.01 × 11.6

6. 9.001 × 4.2

7. 3.24 × 0.008

8. 0.012 × 2.9

9. 0.9 × 11.2

10. 0.03 × 4.5

11. 27.8 × 0.023

12. 1.54 × 7.01

Multi-Part Lesson 1

PART E

NAME ______________________ DATE ____________ PERIOD _____

Skills Practice

Multiply Decimals by Decimals

Multiply.

1. 0.3 × 0.5

2. 1.2 × 2.1

3. 2.5 × 6.7

4. 0.4 × 8.3

5. 2.3 × 1.21

6. 0.6 × 0.91

7. 6.5 × 0.04

8. 8.54 × 3.27

9. 5.02 × 1.07

10. 0.003 × 2.9

11. 0.93 × 6.8

12. 7.1 × 0.004

13. 3.007 × 6.1

14. 2.52 × 0.15

15. 2.6 × 5.46

16. 16.25 × 1.3

17. 3.5 × 24.09

18. 0.025 × 17.1

19. 11.04 × 6.18

20. 14.83 × 16.7

21. 27.1 × 10.15

22. 41.2 × 10.34

NAME ______________________ DATE __________ PERIOD _____

Reteach

Estimate Quotients

When estimating a quotient, find numbers that are compatible with each other to make it easier to divide mentally.

Example 1 **Estimate a Quotient**

Estimate. 15.1 ÷ 5.7

$5.7\overline{)15.1} \longrightarrow \begin{array}{r} 3 \\ 5\overline{)15} \end{array}$

Round 5.7 to 5 since 15 and 5 are compatible numbers.

The quotient is *about* 3.

Exercises

Estimate each quotient.

1. 71.4 ÷ 9.3

2. 53.8 ÷ 17.2

3. 12.6 ÷ 4.2

4. 31.5 ÷ 7.9

5. 50.2 ÷ 5.3

6. 18.9 ÷ 5.8

7. 24.7 ÷ 4.8

8. 9.2 ÷ 4.7

9. 34.2 ÷ 4.5

10. 99.1 ÷ 24.7

11. 44.9 ÷ 9.3

12. 19.4 ÷ 2.1

Multi-Part Lesson 2 PART A

NAME ________________ DATE ________ PERIOD ____

Skills Practice

Estimate Quotients

Estimate each quotient.

1. $7.9 \div 2.8$

2. $11.6 \div 3.1$

3. $18.4 \div 6.1$

4. $88.3 \div 11.4$

5. $37.4 \div 9.1$

6. $58.4 \div 7.9$

7. $27.3 \div 8.7$

8. $64.1 \div 9.4$

9. $42.1 \div 6.1$

10. $19.4 \div 4.2$

11. $98.7 \div 11.3$

12. $369.1 \div 6.2$

Use estimation to determine whether each answer is reasonable. If the answer is reasonable, write *yes*. If not, write *no* and provide a reasonable estimate.

13. $37.4 \div 18.8 = 4$

14. $126.2 \div 25.9 = 4.8$

15. $103.8 \div 7.7 = 13.5$

16. $26.8 \div 3.2 = 6$

NAME ____________________ DATE __________ PERIOD _____

Reteach

Divide Decimals by Whole Numbers

When you divide a decimal by a whole number, place the decimal point in the quotient above the decimal point in the dividend. Then divide as you do with whole numbers.

Example 1 **Find 8.73 ÷ 9.**

Estimate: $9 \div 9 = 1$

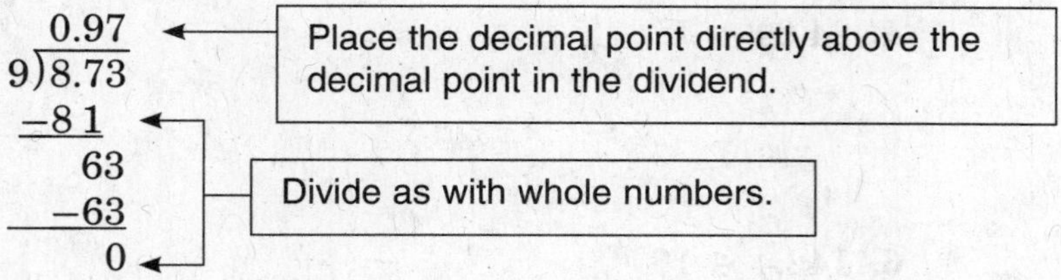

$8.73 \div 9 = 0.97$ Compared with the estimate, the quotient is reasonable.

Example 2 **Find 8.58 ÷ 12.**

Estimate: $10 \div 10 = 1$

```
     0.715   ← Place the decimal point.
12)8.580
  −8 4
     18
    −12       Annex a zero to continue dividing.
     60
    −60
      0
```

$8.58 \div 12 = 0.715$ Compared with the estimate, the quotient is reasonable.

Exercises

Divide.

1. $9.2 \div 4$ **2.** $4.5 \div 5$ **3.** $8.6 \div 2$ **4.** $2.89 \div 4$

5. $3.2 \div 4$ **6.** $7.2 \div 3$ **7.** $7.5 \div 5$ **8.** $3.45 \div 15$

9. $11.8 \div 4$ **10.** $1.09 \div 2$ **11.** $7.6 \div 8$ **12.** $4.56 \div 3$

Multi-Part Lesson 2 PART C

NAME ______________________ DATE __________ PERIOD ______

Skills Practice

Divide Decimals by Whole Numbers

Divide. Round to the nearest tenth if necessary.

1. 9.6 ÷ 3

2. 5.15 ÷ 5

3. 16.08 ÷ 2

4. 24.64 ÷ 7

5. 132.22 ÷ 11

6. 142.4 ÷ 16

7. 79.2 ÷ 9

8. 47.4 ÷ 15

9. 217.14 ÷ 21

10. 34.65 ÷ 5

11. 20.72 ÷ 8

12. 72.6 ÷ 10

13. 57.48 ÷ 15

14. 264.5 ÷ 25

15. 317.59 ÷ 34

16. 122.32 ÷ 11

17. 42.48 ÷ 18

18. 323.31 ÷ 24

NAME ______________________ DATE ____________ PERIOD ______

Reteach

Divide Decimals by Decimals

When you divide a decimal by a decimal, multiply both the divisor and the dividend by the same power of ten. Then divide as with whole numbers.

Example 1 **Find 10.14 ÷ 5.2.**

Estimate: 10 ÷ 5 = 2

Multiply by 10 to make a whole number.

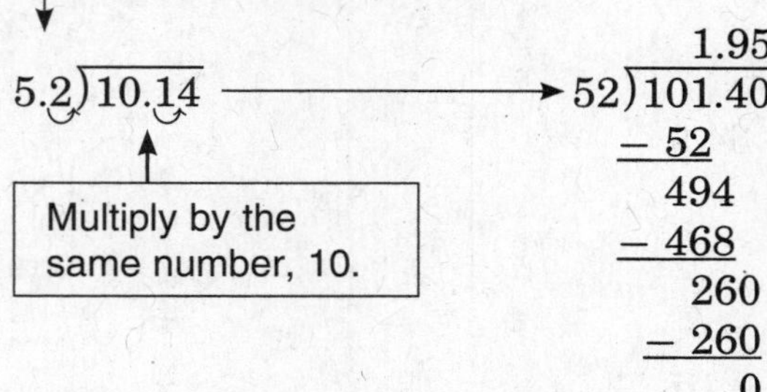

Place the decimal point.
Divide as with whole numbers.

Annex a zero to continue.

10.14 divided by 5.2 is 1.95.

Compare the quotient with the estimate.

Check 1.95 × 5.2 = 10.14 ✓

Example 2 **Find 4.09 ÷ 0.02.**

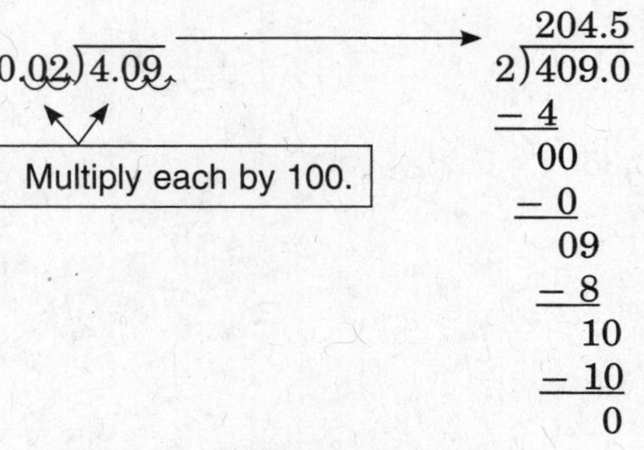

Place the decimal point.
Divide.

Write a zero in the dividend and continue to divide.

4.09 divided by 0.02 is 204.5.

Check 204.5 × 0.02 = 4.09 ✓

Exercises

Divide.

1. 9.8 ÷ 1.4 **2.** 4.41 ÷ 2.1 **3.** 16.848 ÷ 0.72

4. 8.652 ÷ 1.2 **5.** 0.5 ÷ 0.001 **6.** 9.594 ÷ 0.06

NAME ______________________ DATE __________ PERIOD ____

Skills Practice

Divide Decimals by Decimals

Divide.

1. 4.86 ÷ 0.2

2. 2.52 ÷ 0.7

3. 14.4 ÷ 1.2

4. 17.1 ÷ 3.8

5. 3.96 ÷ 1.32

6. 628.2 ÷ 34.9

7. 0.105 ÷ 0.5

8. 1.296 ÷ 0.16

9. 3.825 ÷ 2.5

10. 8.253 ÷ 0.5

11. 0.9944 ÷ 0.8

12. 1.638 ÷ 0.35

13. 13.59 ÷ 0.75

14. 4.4208 ÷ 1.8

15. 16.16 ÷ 0.2

16. 158.1 ÷ 5.1

17. 247.5 ÷ 3.3

18. 0.132 ÷ 1.1

NAME ______________________ DATE ____________ PERIOD _____

Reteach

Multiply by Powers of 10

- **When multiplying a decimal by a power of 10 greater than 1,** move the decimal point to the ***right*** the same number of places as the number of zeros in the power of 10.
- **When multiplying a decimal by a power of 10 less than 1,** move the decimal point to the ***left*** the same number of places as there are places after the decimal point in the power of 10.

Example 1 **Find 7.24 × 1,000.**

$7.24 \times 1{,}000 = \mathbf{7.240}$ — 1,000 has 3 zeros, so move the decimal point 3 places to the right. Annex zeros as needed.

$= 7{,}240$ — Remove the decimal point and add a comma.

Example 2 **Find 36.4 × 0.001.**

$36.4 \times 0.001 = \mathbf{036.4}$ — There are 3 places after the decimal point in 0.001, so move the decimal point 3 places to the left. Annex zeros as needed.

$= 0.0364$ — Annex a zero in front of the decimal point.

Exercises

Find each product.

1. 4.5 × 100

2. 0.298 × 1,000

3. 6.31 × 10

4. 0.34 × 1,000

5. 8.1 × 10,000

6. 44.73 × 10

7. 52.1 × 0.01

8. 12.21 × 0.001

9. 0.56 × 0.1

10. 12.8 × 0.001

11. 35.2 × 0.1

12. 0.7 × 0.01

NAME ______________________ DATE ____________ PERIOD ______

Skills Practice

Multiply by Powers of 10

Find each product.

1. 0.37 × 1,000

2. 14.75 × 100

3. 8.92 × 1,000

4. 9.267 × 100

5. 4.365 × 0.01

6. 67.91 × 0.1

7. 71.23 × 0.001

8. 523.9 × 10

9. 0.0497 × 10

10. 27.37 × 0.01

11. 10.39 × 1,000

12. 78.01 × 0.001

13. 0.975 × 100

14. 9,810 × 0.1

15. 24.98 × 1,000

NAME ______________________ DATE __________ PERIOD _____

Reteach

Divide by Powers of 10

- **When dividing a decimal by a power of 10 greater than 1,** move the decimal point to the ***left*** the same number of places as the number of zeros in the power of 10.
- **When dividing a decimal by a power of 10 less than 1,** move the decimal point to the ***right*** the same number of places as there are places after the decimal point in the power of 10.

Example 1 **Find 13.4 ÷ 100.**

$13.4 \div 100 = 13.4$ — 100 has two zeros, so move the decimal point 2 places to the left.

$= 0.134$ — Since all digits are decimal digits, insert a leading zero.

Example 2 **Find 0.67 ÷ 0.01.**

$0.67 \div 0.01 = 0.67$ — 0.01 has two places after the decimal point, so move the decimal point 2 places to the right.

$= 67$ — Remove the decimal point.

Exercises

Find each quotient.

1. 7.586 ÷ 1,000 **2.** 243 ÷ 100 **3.** 10.9 ÷ 10

4. 16.82 ÷ 100 **5.** 0.95 ÷ 1,000 **6.** 1.536 ÷ 10

7. 5.86 ÷ 0.01 **8.** 37.21÷ 0.1 **9.** 6.03 ÷ 0.001

10. 0.284 ÷ 0.1 **11.** 84 ÷ 0.001 **12.** 41.4 ÷ 0.01

Multi-Part Lesson 3 PART C

NAME ______________________ DATE ____________ PERIOD _____

Skills Practice

Divide by Powers of 10

Find each quotient.

1. 3.94 ÷ 1,000

2. 28.7 ÷ 100

3. 543 ÷ 10

4. 19.6 ÷ 100

5. 0.453 ÷ 1,000

6. 268 ÷ 10

7. 0.76 ÷ 0.01

8. 82 ÷ 0.1

9. 34.5 ÷ 0.001

10. 1.392 ÷ 0.1

11. 14.4 ÷ 0.001

12. 2.03 ÷ 0.01

13. 0.125 ÷ 0.1

14. 71 ÷ 100

15. 0.88 ÷ 1,000

NAME ________________ DATE ________ PERIOD ____

Reteach

Problem-Solving Investigation: Reasonable Answers

When solving problems, one strategy that is helpful is to *determine reasonable answers*. If you are solving a problem with large values or a problem with information that you are unfamiliar with, it may be helpful to look back at your answer to determine if it is reasonable.

You can use the *determine reasonable answers* strategy, along with the following four-step problem-solving plan, to solve a problem.

1 Understand – Read and get a general understanding of the problem.

2 Plan – Make a plan to solve the problem and estimate the solution.

3 Solve – Use your plan to solve the problem.

4 Check – Check the reasonableness of your solution.

Example ANIMALS **The average height of a male chimpanzee is 1.2 meters, and the average height of a female chimpanzee is 1.1 meters. What is a reasonable height in feet of a male chimpanzee? One meter is about the same as one yard.**

Understand We know the average height in meters of a male chimpanzee. We need to find a reasonable height in feet.

Plan One meter is very close to one yard. One yard is equal to 3 feet. So, estimate how many feet would be in 1.2 yards.

Solve 1.2 yards would be more than 3 feet, but less than 6 feet. So, a reasonable average height of a male chimpanzee is about 4 feet.

Check Since 1.2 yd = 3.6 ft, the answer of 4 feet is reasonable.

Exercises

1. **SHOPPING** Alexis wants to buy 2 bracelets for $6.95 each, 1 pair of earrings for $4.99, and 2 necklaces for $8.95 each. Does she need $40 or will $35 be enough? Explain.

2. **FOOD DRIVE** Iyoka's class has a goal of collecting 100 cans of food for a food drive. The five rows of students have collected 15, 31, 22, 29, and 11 cans of food. Has the class met its goal yet? Explain.

NAME ______________________ DATE ____________ PERIOD ______

Multi-Part Lesson 3 PART D

Skills Practice

Problem-Solving Investigation: Reasonable Answers

Use the *determine reasonable answers* strategy to solve each problem.

1. **ANIMALS** A male African elephant weighs 6.5 tons. What is a reasonable weight in pounds of a male African elephant?

2. **AWARDS** The school auditorium holds 3,600 people. Is it reasonable to offer each of the 627 students five tickets for themselves, family, and friends to attend an awards ceremony? Explain.

3. **POPULATION** Use the graph at the right to determine whether 600, 700, or 800 is a reasonable prediction of the enrollment at Midtown Junior High in 2010.

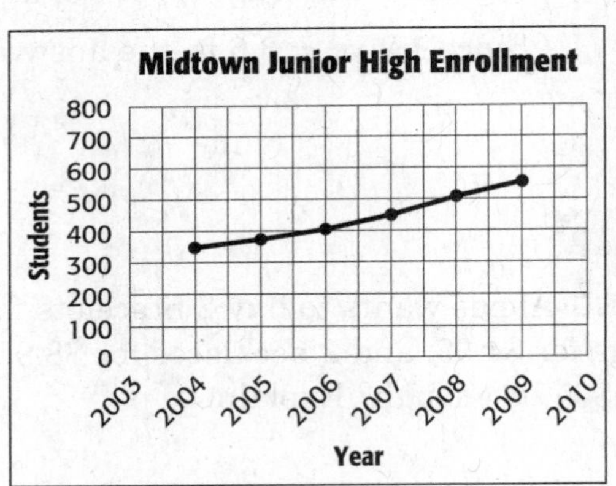

4. **FOOTBALL** This year, 522,530 people attended one team's 8 home pro football games. Which is more reasonable for the number of people that attended each game: 45,000; 55,000; or 65,000?

NAME ______________________ DATE __________ PERIOD _____

Reteach

Estimate Products of Fractions

Numbers that are easy to divide mentally are called **compatible numbers.** One way to estimate products involving fractions is to use compatible numbers.

Example 1 **Estimate $\frac{2}{3} \times 8$.**

Estimate $\frac{2}{3} \times 8$. Make it easier by finding $\frac{1}{3} \times 8$ first.

$\frac{1}{3} \times 9 = ?$ Change 8 to 9 since 3 and 9 are compatible numbers.

$\frac{1}{3} \times 9 = 3$ $\frac{1}{3}$ of 9, or 9 divided by 3, is 3.

$\frac{2}{3} \times 9 = 6$ Since $\frac{1}{3}$ of 9 is 3, $\frac{2}{3}$ of 9 is 2×3 or 6.

So, $\frac{2}{3} \times 8$ is *about* 6.

Another way to estimate products is by rounding fractions to 0, $\frac{1}{2}$, or 1. If the fraction has a numerator much smaller than the denominator, round to 0. If the numerator is about half of the denominator, round to $\frac{1}{2}$. If the numerator and denominator are almost the same, round to 1.

Example 2 **Estimate $\frac{1}{3} \times \frac{5}{6}$.**

$\frac{1}{3} \times \frac{5}{6} \rightarrow \frac{1}{2} \times 1 = \frac{1}{2}$.

So, $\frac{1}{3} \times \frac{5}{6}$ is *about* $\frac{1}{2}$.

You can estimate the product of mixed numbers by rounding to the nearest whole number.

Example 3 **Estimate $3\frac{1}{4} \times 5\frac{7}{8}$.**

Since $3\frac{1}{4}$ rounds to 3 and $5\frac{7}{8}$ rounds to 6, $3\frac{1}{4} \times 5\frac{7}{8} \rightarrow 3 \times 6 = 18$.

So, $3\frac{1}{4} \times 5\frac{7}{8}$ is *about* 18.

Exercises

Estimate each product.

1. $\frac{1}{5} \times 24$

2. $\frac{1}{3}$ of 16

3. $\frac{3}{8} \times 17$

4. $\frac{4}{7}$ of 20

5. $\frac{7}{8} \times \frac{3}{5}$

6. $\frac{11}{12} \times \frac{1}{3}$

7. $\frac{1}{9} \times \frac{1}{12}$

8. $\frac{11}{12} \times \frac{6}{7}$

9. $3\frac{7}{8} \times 10\frac{1}{10}$

10. $2\frac{4}{5} \times 6\frac{1}{12}$

11. $4\frac{7}{8} \times 2\frac{9}{10}$

12. $7\frac{2}{7} \times 5\frac{3}{4}$

NAME ______________________ DATE __________ PERIOD ______

Multi-Part Lesson 1 PART B

Skills Practice

Estimate Products of Fractions

Estimate each product.

1. $\frac{1}{5} \times 26$

2. $\frac{1}{2}$ of 17

3. $\frac{3}{4} \times 35$

4. $\frac{2}{3} \times 35$

5. $\frac{3}{7} \times 29$

6. $\frac{2}{9} \times 26$

7. $\frac{5}{8} \times 41$

8. $\frac{7}{8}$ of 30

9. $\frac{6}{11} \times 32$

10. $\frac{6}{7} \times \frac{1}{8}$

11. $\frac{1}{3} \times \frac{9}{11}$

12. $\frac{10}{11} \times \frac{1}{9}$

13. $\frac{5}{6}$ of $\frac{2}{7}$

14. $\frac{3}{5} \times \frac{6}{7}$

15. $\frac{7}{8} \times \frac{8}{9}$

16. $\frac{5}{9} \times \frac{1}{7}$

17. $\frac{1}{12} \times \frac{5}{9}$

18. $9\frac{1}{8} \times \frac{1}{3}$

19. $2\frac{4}{5} \times 5\frac{1}{4}$

20. $4\frac{1}{3} \times 3\frac{7}{8}$

21. $6\frac{3}{10} \times 4\frac{7}{9}$

22. $3\frac{1}{4} \times 7\frac{7}{8}$

23. $6\frac{7}{12} \times 8\frac{5}{12}$

24. $7\frac{2}{3} \times 9\frac{3}{8}$

25. Estimate $\frac{4}{5}$ of 49.

26. Estimate the product of $2\frac{4}{11}$ and $16\frac{1}{5}$.

NAME ______________________ DATE ____________ PERIOD _____

Reteach

Multiply Fractions and Whole Numbers

You can multiply whole numbers and fractions by writing the whole number as a fraction. Then multiply the numerators and multiply the denominators.

Example 1

Find $6 \times \frac{3}{8}$. Estimate $6 \times \frac{1}{2} = 3$.

$6 \times \frac{3}{8} = \frac{6}{1} \times \frac{3}{8}$ Write 6 as $\frac{6}{1}$.

$= \frac{6 \times 3}{1 \times 8}$ Multiply.

$= \frac{18}{8} = \frac{9}{4}$ or $2\frac{1}{4}$ Simplify. Compare to the estimate.

You can also multiply fractions by using a diagram.

Example 2

Find $\frac{2}{3} \times 4$.

Draw 4 units. Then divide each unit into thirds.

Shade $\frac{2}{3}$ of each unit.

Rearrange to see that $2\frac{2}{3}$ units are shaded.

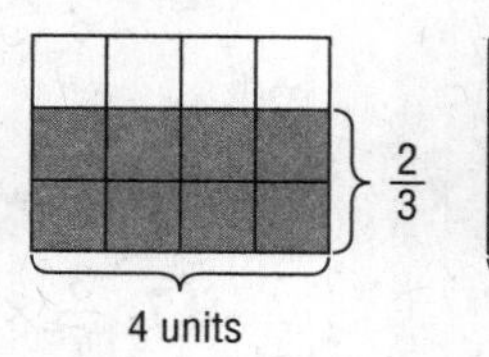

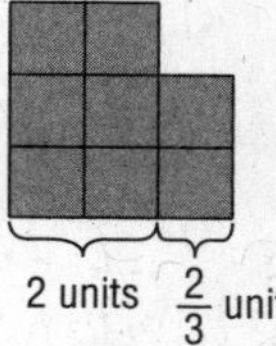

Exercises

Multiply. Write in simplest form.

1. $5 \times \frac{2}{3}$

2. $10 \times \frac{3}{5}$

3. $9 \times \frac{1}{3}$

4. $2 \times \frac{2}{5}$

5. $6 \times \frac{1}{4}$

6. $15 \times \frac{1}{8}$

7. $\frac{2}{3} \times 12$

8. $\frac{4}{5} \times 3$

9. $\frac{4}{5} \times 15$

10. $\frac{1}{6} \times 11$

11. $\frac{2}{7} \times 5$

12. $\frac{5}{6} \times 12$

Multi-Part Lesson 1 PART D

NAME ______________________ DATE ____________ PERIOD _____

Skills Practice

Multiply Fractions and Whole Numbers

1. $10 \times \frac{1}{5}$

2. $12 \times \frac{1}{3}$

3. $18 \times \frac{1}{9}$

4. $15 \times \frac{2}{3}$

5. $16 \times \frac{3}{8}$

6. $8 \times \frac{3}{4}$

7. $7 \times \frac{4}{5}$

8. $10 \times \frac{3}{5}$

9. $18 \times \frac{4}{9}$

10. $20 \times \frac{9}{10}$

11. $16 \times \frac{3}{4}$

12. $14 \times \frac{6}{7}$

13. $\frac{3}{4} \times 12$

14. $\frac{1}{6} \times 9$

15. $\frac{1}{6} \times 18$

16. $\frac{2}{3} \times 18$

17. $\frac{3}{7} \times 14$

18. $\frac{2}{5} \times 20$

19. $\frac{2}{5} \times 12$

20. $\frac{5}{6} \times 10$

21. $\frac{3}{4} \times 15$

22. $\frac{3}{7} \times 8$

23. $\frac{3}{8} \times 9$

24. $\frac{5}{6} \times 13$

25. Find the product of 24 and $\frac{5}{6}$.

26. Find the product of $\frac{7}{8} \times 11$.

NAME _______________ DATE _______ PERIOD _____

Reteach

Problem-Solving Investigation: Draw a Diagram

When solving problems, one strategy that is helpful is to *draw a diagram.* A problem may often describe a situation that is easier to solve visually. You can draw a diagram of the situation and then use the diagram to solve the problem.

You can draw a bar diagram, along with the following four-step problem solving plan, to solve a problem.

1 Understand – Read and get a general understanding of the problem.

2 Plan – Make a plan to solve the problem and estimate the solution.

3 Solve – Use your plan to solve the problem.

4 Check – Check the reasonableness of your solution.

Example 1 **After delivering tomatoes to the neighbors, $\frac{4}{5}$ of all the tomatoes Ryan started with were gone. If he has 4 tomatoes left, how many tomatoes did he deliver?**

Understand You know that Ryan delivered $\frac{4}{5}$ of all the tomatoes. You need to know how many tomatoes he delivered.

Plan Draw a bar diagram.

Solve Draw a bar diagram that represents the amount of tomatoes Ryan started with.

delivered	delivered	delivered	delivered	left
?				4

Determine how many tomatoes were delivered.

4	4	4	4	4
16				4

So, Ryan delivered 16 tomatoes.

Check Check the diagram to make sure that it meets all of the requirements. Since the diagram is correct, the answer is correct.

Exercise

Latasha is gluing red and blue gem stones onto a picture frame. Of the gem stones, $\frac{2}{9}$ are red and 14 are blue. How many are red? Use the *draw a diagram* strategy.

NAME ______________________ DATE __________ PERIOD _____

Multi-Part Lesson 1 PART E

Skills Practice

Problem-Solving Investigation: Draw a Diagram

Solve. Use the *draw a diagram* strategy.

1. **TRAVEL** Jasmine biked $\frac{3}{8}$ the distance from her school to her house. If she has 10 more miles to go, how many miles has she gone?

2. **GARDENING** Ms. Kennedy grew $\frac{3}{5}$ as many zinnias as Mr. Hadam. If Ms. Kennedy has 12 zinnias, how many zinnias do they have in all?

3. **SPORTS** Benny plays two sports every afternoon. He plays soccer for $\frac{3}{4}$ of that time and plays baseball for 30 minutes. How many more minutes does he play soccer than baseball?

Multi-Part Lesson 2 PART D

NAME ______________________ DATE __________ PERIOD _____

Reteach

Multiply Mixed Numbers

To multiply mixed numbers, write the mixed numbers as improper fractions and then multiply as with fractions.

Example 1 **Find $\frac{1}{4} \times 1\frac{2}{3}$.** **Estimate. Use compatible numbers. $\frac{1}{2} \times 2 = 1$**

$\frac{1}{4} \times 1\frac{2}{3} = \frac{1}{4} \times \frac{5}{3}$ Write $1\frac{2}{3}$ as $\frac{5}{3}$.

$= \frac{1 \times 5}{4 \times 3}$ Multiply.

$= \frac{5}{12}$ Simplify. Compare to the estimate.

Example 2 **Find $1\frac{1}{3} \times 2\frac{1}{4}$.**

$1\frac{1}{3} \times 2\frac{1}{4} = \frac{4}{3} \times \frac{9}{4}$ Convert mixed numbers to improper fractions.

$= \frac{\overset{1}{\cancel{4}}}{\underset{1}{\cancel{3}}} \times \frac{\overset{3}{\cancel{9}}}{\underset{1}{\cancel{4}}}$ Divide the numerator and denominator by their common factors, 3 and 4.

$= \frac{3}{1}$ or 3 Simplify.

Exercises

Multiply. Write in simplest form.

1. $\frac{1}{3} \times 1\frac{1}{3}$ **2.** $1\frac{1}{5} \times \frac{3}{4}$ **3.** $\frac{2}{3} \times 1\frac{3}{5}$ **4.** $\frac{2}{3} \times 3\frac{1}{2}$

5. $\frac{2}{9} \times 1\frac{1}{6}$ **6.** $2\frac{4}{9} \times \frac{4}{11}$ **7.** $2\frac{1}{2} \times 1\frac{1}{3}$ **8.** $1\frac{1}{4} \times 3\frac{3}{5}$

9. $8\frac{1}{5} \times 1\frac{1}{4}$ **10.** $1\frac{3}{8} \times 2\frac{1}{2}$ **11.** $4\frac{2}{3} \times 1\frac{1}{8}$ **12.** $1\frac{1}{9} \times 3\frac{2}{5}$

13. Find the product of $\frac{1}{5}$ and $3\frac{1}{3}$.

14. Simplify $4\frac{2}{3} \times 1\frac{1}{4}$.

NAME ____________________ DATE __________ PERIOD _____

Multi-Part Lesson 2 PART D

Skills Practice

Multiply Mixed Numbers

Multiply. Write in simplest form.

1. $\frac{1}{3} \times 1\frac{1}{4}$

2. $2\frac{1}{2} \times \frac{3}{5}$

3. $\frac{3}{4} \times 3\frac{1}{3}$

4. $6\frac{1}{5} \times \frac{1}{2}$

5. $\frac{5}{7} \times 4\frac{1}{5}$

6. $\frac{4}{7} \times 3\frac{1}{9}$

7. $4\frac{1}{6} \times \frac{9}{10}$

8. $\frac{8}{9} \times 5\frac{1}{7}$

9. $\frac{5}{7} \times 4\frac{3}{8}$

10. $2\frac{4}{9} \times \frac{6}{11}$

11. $2\frac{5}{8} \times \frac{1}{6}$

12. $\frac{2}{5} \times 1\frac{2}{5}$

13. $1\frac{3}{5} \times 3\frac{2}{3}$

14. $1\frac{3}{8} \times 2\frac{2}{7}$

15. $3\frac{1}{3} \times 2\frac{1}{4}$

16. $3\frac{3}{4} \times 2\frac{4}{5}$

17. $5\frac{3}{4} \times 1\frac{1}{11}$

18. $2\frac{5}{8} \times 2\frac{5}{7}$

19. $2\frac{2}{9} \times 4\frac{4}{5}$

20. $5\frac{3}{8} \times 2\frac{2}{7}$

21. $6\frac{2}{3} \times 5\frac{2}{11}$

22. $6\frac{2}{5} \times 5\frac{5}{9}$

23. $8\frac{8}{9} \times 3\frac{1}{10}$

24. $9\frac{3}{5} \times 8\frac{7}{8}$

25. Find the product of $\frac{2}{3} \times 5\frac{1}{6}$.

26. Simplify $4\frac{1}{2} \times 6\frac{2}{3}$.

NAME ______________________ DATE ____________ PERIOD _____

Reteach

Divide Whole Numbers by Fractions

When the product of two numbers is 1, the numbers are called reciprocals.

Example 1 **Find the reciprocal of $\frac{5}{9}$.**

Since $\frac{5}{9} \times \frac{9}{5} = 1$, the reciprocal of $\frac{5}{9}$ is $\frac{9}{5}$.

Example 2 **Find the reciprocal of 8.**

Since $8 \times \frac{1}{8} = 1$, the reciprocal of 8 is $\frac{1}{8}$.

You can use reciprocals to divide whole numbers by fractions. To divide by a fraction, multiply by its reciprocal.

Example 3 **Find $4 \div \frac{1}{3}$.**

$4 \div \frac{1}{3} = \frac{4}{1} \times \frac{3}{1}$ Multiply by the reciprocal, $\frac{3}{1}$.

$= \frac{12}{1}$ or 12 Simplify.

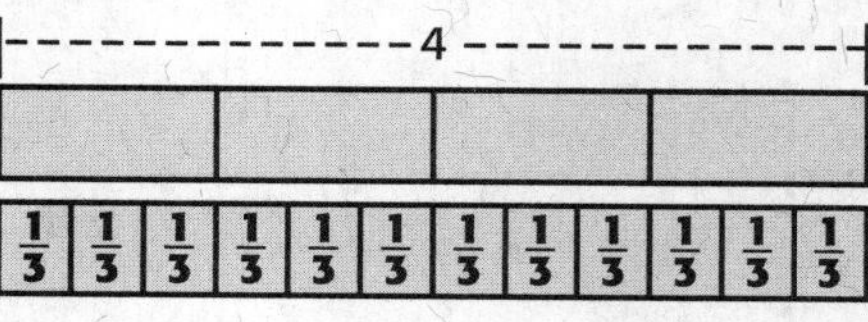

Exercises

Find the reciprocal of each number.

1. $\frac{1}{2}$ **2.** $\frac{1}{6}$ **3.** $\frac{4}{11}$ **4.** $\frac{3}{5}$

Divide. Write in simplest form.

5. $3 \div \frac{2}{5}$ **6.** $9 \div \frac{1}{2}$ **7.** $2 \div \frac{1}{4}$ **8.** $1 \div \frac{3}{4}$

9. $4 \div \frac{1}{2}$ **10.** $5 \div \frac{1}{10}$ **11.** $12 \div \frac{5}{6}$ **12.** $9 \div \frac{2}{3}$

13. $4 \div \frac{7}{12}$ **14.** $10 \div \frac{8}{9}$ **15.** $3 \div \frac{5}{8}$ **16.** $4 \div \frac{7}{9}$

Multi-Part Lesson 3 PART B

NAME ________________________ DATE ____________ PERIOD _____

Skills Practice

Divide Whole Numbers by Fractions

Find the reciprocal of each number.

1. $\frac{1}{2}$

2. $\frac{3}{5}$

3. $\frac{4}{7}$

4. $\frac{8}{11}$

5. $\frac{7}{12}$

6. $\frac{9}{10}$

7. $\frac{5}{8}$

8. $\frac{3}{10}$

Divide. Write in simplest form.

9. $2 \div \frac{1}{3}$

10. $4 \div \frac{1}{2}$

11. $1 \div \frac{3}{5}$

12. $8 \div \frac{4}{5}$

13. $7 \div \frac{5}{6}$

14. $10 \div \frac{1}{4}$

15. $9 \div \frac{3}{8}$

16. $9 \div \frac{3}{4}$

17. $2 \div \frac{4}{7}$

18. $15 \div \frac{5}{9}$

19. $6 \div \frac{3}{11}$

20. $9 \div \frac{5}{12}$

21. $6 \div \frac{5}{12}$

22. $5 \div \frac{10}{11}$

23. $9 \div \frac{1}{7}$

24. $7 \div \frac{8}{9}$

25. $5 \div \frac{9}{11}$

26. $5 \div \frac{4}{9}$

27. Simplify $18 \div \frac{2}{9}$.

28. Simplify $21 \div \frac{7}{9}$.

NAME ______________________ DATE ____________ PERIOD _____

Reteach

Divide Mixed Numbers

To divide mixed numbers, express each mixed number as an improper fraction. Then divide as with fractions.

Example 1 **Find $1\frac{2}{3} \div \frac{3}{4}$.**

$1\frac{2}{3} \div \frac{3}{4} = \frac{5}{3} \div \frac{3}{4}$ — Write the mixed number as an improper fraction.

$= \frac{5}{3} \times \frac{4}{3}$ — Multiply by the reciprocal.

$= \frac{20}{9}$ or $2\frac{2}{9}$ — Simplify.

Example 2 **Find $2\frac{2}{3} \div 1\frac{1}{5}$.** — **Estimate:** 3 ÷ 1 = 3

$2\frac{2}{3} \div 1\frac{1}{5} = \frac{8}{3} \div \frac{6}{5}$ — Write mixed numbers as improper fractions.

$= \frac{8}{3} \times \frac{5}{6}$ — Multiply by the reciprocal, $\frac{5}{6}$.

$= \frac{\overset{4}{\cancel{8}} \times 5}{3 \times \underset{3}{\cancel{6}}}$ — Divide 8 and 6 by the GCF, 2.

$= \frac{20}{9}$ or $2\frac{2}{9}$ — Simplify. Compare to the estimate.

Exercises

Divide. Write in simplest form.

1. $2\frac{1}{2} \div \frac{4}{5}$ **2.** $9 \div 1\frac{1}{9}$ **3.** $5 \div 1\frac{3}{7}$ **4.** $2\frac{1}{3} \div \frac{7}{9}$

5. $5\frac{2}{5} \div \frac{9}{10}$ **6.** $2\frac{1}{4} \div \frac{2}{7}$ **7.** $2\frac{1}{2} \div 3\frac{1}{3}$ **8.** $7\frac{1}{2} \div 1\frac{2}{3}$

9. $1\frac{2}{3} \div 1\frac{1}{4}$ **10.** $4\frac{4}{5} \div 2\frac{6}{7}$ **11.** $5\frac{1}{10} \div 1\frac{8}{9}$ **12.** $2\frac{3}{8} \div 2\frac{1}{4}$

13. Simplify $6 \div 4\frac{3}{5}$.

14. Simplify $4\frac{2}{3} \div 1\frac{3}{4}$.

Multi-Part Lesson 3 PART E

NAME ______________________ DATE ____________ PERIOD _____

Skills Practice

Divide Mixed Numbers

Divide. Write in simplest form.

1. $2\frac{5}{6} \div 6\frac{4}{5}$

2. $4\frac{6}{7} \div 3\frac{2}{5}$

3. $31\frac{2}{3} \div 7\frac{3}{5}$

4. $3 \div 1\frac{1}{3}$

5. $6 \div 2\frac{2}{5}$

6. $1\frac{3}{4} \div \frac{3}{4}$

7. $2 \div 4\frac{2}{7}$

8. $7 \div 3\frac{1}{9}$

9. $6\frac{2}{3} \div \frac{4}{5}$

10. $1\frac{2}{9} \div \frac{5}{6}$

11. $6 \div 1\frac{7}{20}$

12. $\frac{7}{10} \div 2\frac{5}{8}$

13. $3\frac{5}{6} \div 1\frac{1}{3}$

14. $1\frac{7}{9} \div \frac{4}{9}$

15. $5 \div 8\frac{3}{4}$

16. $2\frac{2}{9} \div 1\frac{1}{3}$

17. $3\frac{1}{5} \div 1\frac{7}{9}$

18. $6 \div 3\frac{1}{3}$

19. $3\frac{2}{3} \div 2\frac{2}{3}$

20. $4\frac{1}{4} \div 2\frac{5}{8}$

21. $4\frac{1}{3} \div 3\frac{1}{3}$

22. $4\frac{2}{3} \div 2\frac{2}{9}$

23. $6\frac{3}{5} \div 2\frac{3}{5}$

24. $5\frac{5}{8} \div 3\frac{3}{4}$

25. Simplify $10\frac{3}{4} \div 6\frac{1}{2}$.

26. Simplify $9\frac{4}{9} \div \frac{4}{9}$.

NAME ______________________ DATE __________ PERIOD _____

Multi-Part Lesson 1 PART B

Reteach

Ratios

A **ratio** is a comparison of two numbers by division. A common way to express a ratio is as a fraction in simplest form. Ratios can also be written in other ways. For example, the ratio $\frac{2}{3}$ can be written as 2 to 3, 2 out of 3, or 2:3.

Examples **Refer to the diagram at the right.**

1 Write the ratio in simplest form that compares the number of circles to the number of triangles.

circles → 4 / triangles → 5 $\frac{4}{5}$ The GCF of 4 and 5 is 1.

So, the ratio of circles to triangles is $\frac{4}{5}$, 4 to 5, or 4:5.

For every 4 circles, there are 5 triangles.

2 Write the ratio in simplest form that compares the number of circles to the total number of figures.

circles → 4 / total figures → 10: $\frac{4}{10} \underset{\div 2}{\overset{\div 2}{=}} \frac{2}{5}$ The GCF of 4 and 10 is 2.

The ratio of circles to the total number of figures is $\frac{2}{5}$, 2 to 5, or 2:5. For every two circles, there are five total figures.

Example 3 **Divide 24 roses into two groups so the ratio is 3 to 5.**

Use a bar diagram. Show a group of 3 and a group of 5.

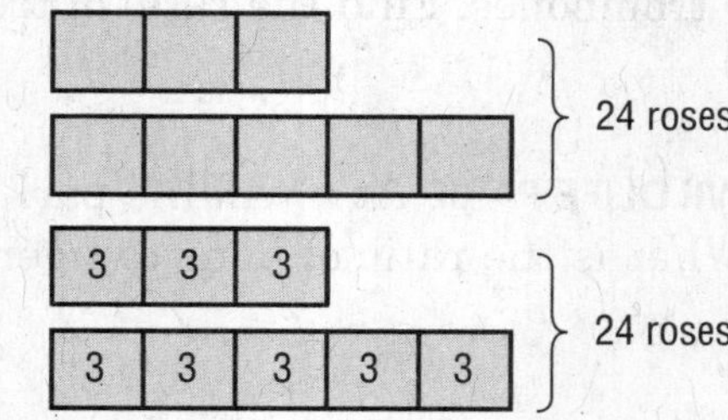

Because there are 8 sections, each section represents 24 ÷ 8, or 3 roses.

There are 9 roses in the first group and 15 roses in the second group.

Exercises

Write each ratio as a fraction in simplest form. Then explain its meaning.

1. 2 guppies to 6 seahorses

2. 12 puppies to 15 kittens

3. **SPELLING** A sentence has 5 misspelled words and 15 correctly spelled words. Find the ratio of misspelled words to correctly spelled words.

Multi-Part Lesson 1 PART B

NAME ______________________ DATE ____________ PERIOD _____

Skills Practice

Ratios

For Exercises 1–10, write each ratio as a fraction in simplest form. Then explain its meaning.

1. 3 sailboats to 6 fan boats

2. 4 tulips to 6 daffodils

3. 5 ducks to 30 geese

4. 5 baseballs to 25 softballs

5. 6 poodles to 18 beagles

6. 10 brown eggs to 12 white eggs

7. **WALLPAPER** The design on Santana's wall includes 16 pink stripes and 20 green stripes. Find the ratio of pink stripes to green stripes.

8. **JAZZ BAND** In the jazz band at Wyatt's school, there are 15 trumpets and 9 trombones. Find the ratio of trombones to trumpets.

9. **WILDLIFE PARK** At a wildlife park, Zoey counted 10 lions and 14 tigers. What is the ratio of lions to tigers?

10. **MAILBOX** In one week, Latrina received 18 letters and 8 bills. What was the ratio of bills to letters?

11. **GYM CLASS** Mr. Riley allowed a class of sixth graders to choose an activity for their gym class. What is the ratio of students playing volleyball to the total number of students? Write the ratio as a fraction in simplest form. Then explain its meaning.

Gym Class Activities	
Activity	**Students**
Volleyball	12
Softball	21
Soccer	3

12. **FRUIT SALAD** In a fruit salad, there are 12 strawberries, 14 grapes, 6 kiwis, and 4 papayas. Find the ratio of kiwis to the total number of pieces of fruit in the fruit salad. Write the ratio as a fraction in simplest form. Then explain its meaning.

Multi-Part Lesson 1 PART D

NAME ______________________________ DATE ______________ PERIOD ______

Reteach

Rates

A **rate** is a ratio of two measurements having different kinds of units. When a rate is simplified so that it has a denominator of 1, it is called a **unit rate.**

Example 1

Use a bar diagram to show the ratio *20 students to 5 computers* as a unit rate.

20 students

1 computer	1 computer	1 computer	1 computer	1 computer

4 students

The bar diagram shows the number of students divided by the number of computers. It represents the number of students per computer.

The ratio written as an unit rate is *4 students to 1 computer*.

You can also find a unit rate by dividing.

Example 2

Benito ate 48 raisins in 8 minutes. How many raisins did he eat per minute, if he ate the same number each minute?

$$\frac{48 \text{ raisins}}{8 \text{ minutes}} = \frac{6 \text{ raisins}}{1 \text{ minute}} \quad (\div 8 \text{ numerator and denominator})$$

Divide the numerator and denominator by 8 to get a denominator of 1.

The unit rate is 6 raisins per minute.

Exercises

Write each rate as a unit rate.

1. 6 eggs for 3 people

2. $12 for 4 pounds

3. 40 pages in 8 days

4. **GROCERIES** Mr. Gonzalez spends $135 for 5 bags of groceries. How much does he spend per bag of groceries, if each bag costs the same?

5. **TRAIN** Ms. Terry travels by train to see famous theme parks. She travels a distance of 728 miles in 8 hours. If the train maintains a constant speed, how many miles does she travel in one hour?

6. **FOOTBALL** A quarterback throws 222 yards in 6 games. How many yards does he throw in one game if he throws the same amount in each game?

NAME ______________________ DATE ____________ PERIOD _____

Skills Practice

Rates

Write each rate as a unit rate.

1. 14 hours in 2 weeks

2. 36 pieces of candy for 6 children

3. 8 teaspoons for 4 cups

4. 8 tomatoes for $2

5. $28 for 4 hours

6. 150 miles in 3 hours

7. $18 for 3 CDs

8. 48 logs on 6 trucks

9. Write the ratio *$12 dollars for 3 tickets* as a unit rate.

10. CHORES Wayne raked 30 bags of leaves in 3 hours. If he raked the same number of bags each hour, how many bags of leaves did he rake in one hour?

11. QUIZZES Mr. Ordonez gives his math students 34 quizzes during 17 weeks of school. If he gave the same number of quizzes each week, how many quizzes does Mr. Ordonez give his students every week?

12. TOURS It cost Mrs. Sapanaro $245 for her and 6 people to take a day-long guided tour of the Everglades. How much does the guided tour cost per person?

13. RUNNING Stephanie ran 1 lap in 6 minutes. At this rate, how far would she run in 30 minutes?

14. ALTITUDE In general, the air temperature decreases 12°F for every 4,000 feet increase in altitude. If a hiker climbs 3,000 feet, by how much can she expect the temperature to decrease?

15. PURCHASES One bottle of shampoo costs $6 for 8 ounces. A second bottle costs $4 for 5 ounces of shampoo. Which has the lower unit rate? How much lower?

NAME ______________________ DATE __________ PERIOD ______

Reteach

Ratio Tables

A **ratio table** organizes data into columns that are filled with pairs of numbers that have the same ratio, or are equivalent. **Equivalent ratios** express the same relationship between two quantities.

Example 1 BAKING **You need 1 cup of rolled oats to make 24 oatmeal cookies. Use the ratio table below to find how many oatmeal cookies you can make with 5 cups of rolled oats.**

Cups of Oats	1				5
Oatmeal Cookies	24				■

Find a pattern and extend it.

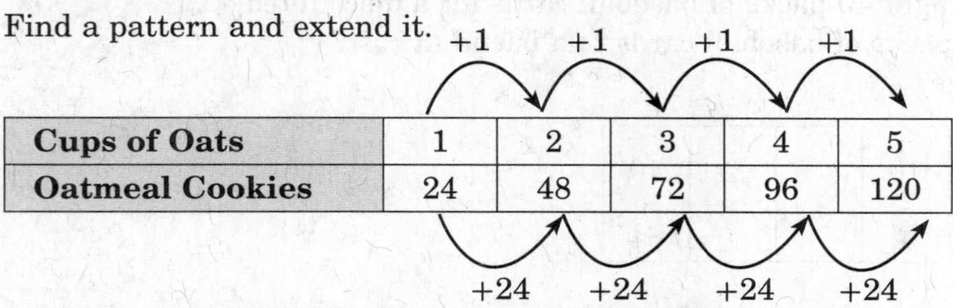

Cups of Oats	1	2	3	4	5
Oatmeal Cookies	24	48	72	96	120

So, 120 oatmeal cookies can be made with 5 cups of rolled oats.

Multiplying or dividing two related quanitities by the same number is called **scaling.** You may sometimes need to *scale back* and then *scale forward* or vice versa to find an equivalent ratio.

Example 2 SHOPPING **A department store has socks on sale for 4 pairs for $10. Use the ratio table at the right to find the cost of 6 pairs of socks.**

Pairs of Socks		4	6
Cost in Dollars		10	■

There is no whole number by which you can multiply 4 to get 6. Instead, scale back to 2 and then forward to 6.

Pairs of Socks	2	4	6
Cost in Dollars	5	10	15

÷ 2 × 3

So, the cost of 6 pairs of socks would be $15.

Exercises

For Exercises 1–2, use the ratio tables given to solve each problem.

1. **EXERCISE** Keewan bikes 6 miles in 30 minutes. At this rate, how long would it take him to bike 18 miles?

Distance Biked (mi)	6		18
Time (min)	30		■

2. **HOBBIES** Christine is making fleece blankets. 6 yards of fleece will make 2 blankets. How many blankets can she make with 9 yards of fleece?

Yards of Fleece		6	9
Number of Blankets		2	■

NAME ______________________ DATE __________ PERIOD _____

Skills Practice

Ratio Tables

Use the ratio table given to solve each problem.

1. **BAKING** A recipe for 1 apple pie calls for 6 cups of sliced apples. How many cups of sliced apples are needed to make 4 apple pies?

Number of Pies	1			4
Cups of Sliced Apples	6			■

2. **BASEBALL CARDS** Justin bought 40 packs of baseball cards for a discounted price of $64. If he sells 10 packs of baseball cards to a friend at cost, how much should he charge?

Number of Baseball Card Packs	10			40
Cost in Dollars	■			64

3. **SOUP** A recipe that yields 12 cups of soup calls for 28 ounces of beef broth. How many ounces of beef broth do you need to make 18 cups of the soup?

Number of Cups		12	18
Ounces of Beef Broth		28	■

4. **ANIMALS** At a dog shelter, a 24-pound bag of dog food will feed 36 dogs a day. How many dogs would you expect to feed with a 16-pound bag of dog food?

Pounds of Dog Food	16	24	
Number of Dogs Fed	■	36	

5. **AUTOMOBILES** Mr. Fink's economy car can travel 420 miles on a 12-gallon tank of gas. Determine how many miles he can travel on 8 gallons.

Miles	420		■
Gallons	12		8

NAME ______________________ DATE ____________ PERIOD ______

Reteach

Problem-Solving Investigation: Look for a Pattern

When solving problems, one strategy that is helpful is to *look for a pattern*. In some problem situations, you can extend and examine a pattern in order to solve the problem.

You can use the *look for a pattern* strategy, along with the following four-step problem solving plan to solve a problem.

1 Understand – Read and get a general understanding of the problem.

2 Plan – Make a plan to solve the problem and estimate the solution.

3 Solve – Use your plan to solve the problem.

4 Check – Check the reasonableness of your solution.

Example MEDICINE **Monisha has the flu. The doctor gave her medicine to take over the next 2 weeks. The first 3 days she is to take 2 pills a day. Then the remaining days she is to take 1 pill. How many pills will Monisha have taken at the end of the 2 weeks?**

Understand You know she is to take the medicine for 2 weeks. You also know she is to take 2 pills the first 3 days and then only 1 pill the remaining days. You need to find the total number of pills.

Plan Start with the first week and look for a pattern.

Solve

Day	1	2	3	4	5	6	7
Number of Pills	2	2	2	1	1	1	1
Total Pills	2	2 + 2 = 4	4 + 2 = 6	6 + 1 = 7	7 + 1 = 8	8 + 1 = 9	9 + 1 = 10

After the first few days the number of pills increases by 1. You can add 7 more pills to the total for the first week, 10 + 7 = 17. So, by the end of the 2 weeks, Monisha will have taken 17 pills to get over the flu.

Check You can extend the table for the next 7 days to check the answer.

Exercise

Use the *look for a pattern* strategy to solve.

TIME Buses arrive every 30 minutes at the bus stop. The first bus arrives at 6:20 A.M. Hogan wants to get on the first bus after 8:00 A.M. What time will the bus that Hogan wants to take arrive at the bus stop?

Multi-Part Lesson 2 PART C

NAME ______________________ DATE ____________ PERIOD ______

Skills Practice

Problem-Solving Investigation: Look for a Pattern

Solve. Use the *look for a pattern* strategy.

1. **NUMBER SENSE** Describe the pattern below. Then find the missing number.
1, 20, 400, ________, 160,000

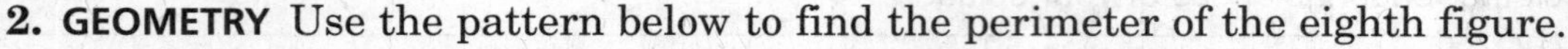

2. **GEOMETRY** Use the pattern below to find the perimeter of the eighth figure.

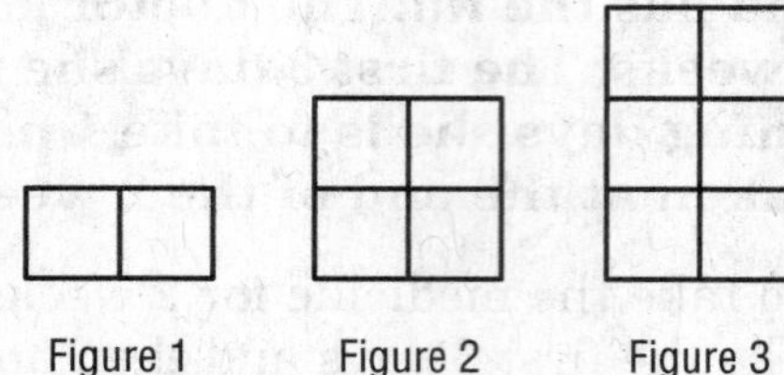

3. **PHYSICAL SCIENCE** A cup of marbles hangs from a rubber band. The length of the rubber band is measured as shown in the graph below. Predict the approximate length of the rubber band if 6 marbles are in the cup.

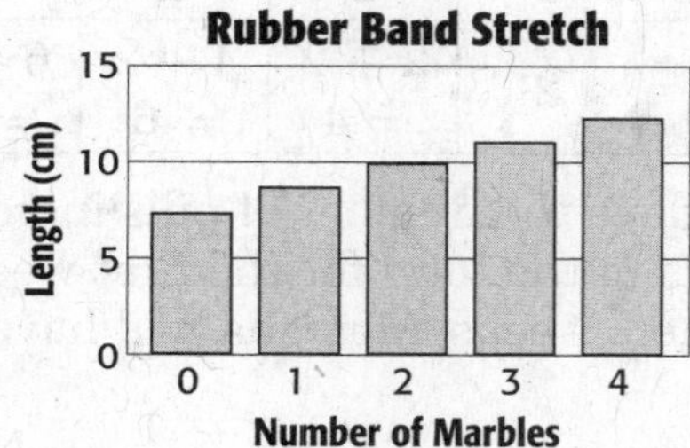

4. **ALLOWANCE** In 2008, Bushra earned $200 in allowance, and Huma earned $150 in allowance. Each year Huma earned $20 more in allowance, and Bushra earned $10 more. In what year will they earn the same amount of money? How much will it be?

NAME ____________________ DATE __________ PERIOD ____

Reteach

Equivalent Ratios

Two ratios are said to be **equivalent ratios** if they have the same unit rate.

Example 1 **Determine if each pair of rates are equivalent. Explain your reasoning.**

\$35 for 7 balls of yarn; \$24 for 4 balls of yarn.

Write each rate as a fraction. Then find its unit rate.

$$\frac{\$35}{7 \text{ balls of yarn}} = \frac{\$5}{1 \text{ ball of yarn}} \quad (\div 7) \qquad \frac{\$24}{4 \text{ balls of yarn}} = \frac{\$6}{1 \text{ ball of yarn}} \quad (\div 4)$$

Since the rates do not share the same unit rate, they are not equivalent.

Example 2 **Determine if each pair of ratios are equivalent. Explain your reasoning.**

8 boys out of 24 students; 4 boys out of 12 students

Write each ratio as a fraction.

$$\frac{8 \text{ boys}}{24 \text{ students}} = \frac{4 \text{ boys}}{12 \text{ students}} \quad (\div 2)$$

← The numerator and the denominator are divided by the same number.

Since the fractions are equivalent, the ratios are equivalent.

Exercises

Determine if each pair of ratios or rates are equivalent. Explain your reasoning.

1. \$12 saved after 2 weeks; \$36 saved after 6 weeks

2. \$9 for 3 magazines; \$20 for 5 magazines

3. 135 miles driven in 3 hours; 225 miles driven in 5 hours

4. 24 computers for 30 students; 48 computers for 70 students

NAME ____________ DATE ________ PERIOD ____

Skills Practice

Equivalent Ratios

Determine if each pair of ratios or rates are equivalent. Explain your reasoning.

1. $18 for 3 bracelets; $30 for 5 bracelets

2. 120 Calories in 2 servings; 360 Calories in 6 servings

3. 4 hours worked for $12; 7 hours worked for $28

4. 15 blank CDs for $5; 45 blank CDs for $15

5. 24 points scored in 4 games; 48 points scored in 10 games

6. 15 out of 20 students own hand-held games; 105 out of 160 students own hand-held games.

7. 30 minutes to jog 3 miles; 50 minutes to jog 5 miles

8. $3 for 6 muffins; $9 for 18 muffins

9. 360 miles driven on 12 gallons of fuel; 270 miles driven on 9 gallons of fuel

10. **SHOPPING** Miguel bought 2 pairs of jeans for $50, and Han bought 4 pairs of jeans for $90. Did they pay the same rate? Explain your reasoning.

NAME ______________________ DATE ____________ PERIOD ____

Reteach

Ratio and Rate Problems

You can solve rate and ratio problems by using a bar diagram or by using a unit rate.

Example 1 NUTRITION **Three servings of broccoli contain 150 calories. How many calories will 5 servings contain?**

Method 1 Use a bar diagram.

Draw a bar diagram to represent the situation.

50	50	50			150 Calories
50	50	50	50	50	? Calories

Each section represents 150 ÷ 3, or 50 Calories.

So, 5 servings of broccoli contain 250 Calories.

Method 2 Use a unit rate.

Step 1 Find the unit rate. $\dfrac{150 \text{ Calories}}{3 \text{ servings}} = \dfrac{\blacksquare \text{ Calories}}{1 \text{ serving}}$ (÷ 3, ÷ 3) $\quad \dfrac{150 \text{ Calories}}{3 \text{ servings}} = \dfrac{50 \text{ Calories}}{1 \text{ serving}}$

Step 2 Multiply. $\dfrac{50 \text{ Calories}}{1 \text{ serving}} \times 5 \text{ servings} = 250 \text{ Calories}$

You can also solve ratio and rate problems by using equivalent fractions.

Example 2 SURVEY **In one survey, three out of five students agreed that the school needs a new cafeteria. Predict how many of the 600 students in the school would agree that the school needs a new cafeteria.**

agree → $\dfrac{3}{5} = \dfrac{\blacksquare}{600}$ ← agree; total → 5, 600 ← total. Write a ratio comparing the number of students who agree to the total number of students.

$\dfrac{3}{5} = \dfrac{360}{600}$ (×120, ×120) Since 5 × 120 = 600, multiply 3 by 120.

So, 360 students would agree that the school needs a new cafeteria.

Exercises

Solve.

1. **MUSIC** Jeremy spent $33 on 3 CDs. At this rate, how much would 5 CDs cost?

2. **AQUARIUM** At an aquarium, 6 out of 18 deliveries are plants. Out of 15 deliveries in one week, how many are plants?

3. **ELECTIONS** Three out of four students surveyed in a school said they will vote for Nuncio for class president. Predict how many of the 340 students in the school would vote for Nuncio.

NAME ______________________ DATE ____________ PERIOD _____

Multi-Part Lesson 3 PART C

Skills Practice

Ratio and Rate Problems

Solve.

1. **GUACAMOLE** Eli is making guacamole. He uses 2 tablespoons of cilantro for every 3 avocados. At this rate, how many tablespoons of cilantro will he need for 9 avocados?

2. **MARBLES** The ratio of blue marbles to white marbles in a bag is 4 to 5. At this rate, how many blue marbles are there if there are 15 white marbles?

3. **FERTILIZER** Ellie must mix 6 tablespoons of plant food for every 2 gallons of water. If she has 6 gallons of water, how much plant food should she use?

4. **STRAWBERRIES** At a local fruit stand, Luisa spends $3.96 for 2 pounds of strawberries. How much can she expect to pay for 4 pounds of strawberries?

5. **POGO STICK** On her pogo stick, Lula made 24 hops in 30 seconds. At this rate, how many hops will she make in 50 seconds?

6. **TESTS** On a test, Matilda answered 12 out of the first 15 problems correctly. If this rate continues, how many of the next 25 problems will she answer correctly?

7. **SOCCER** The Hawks soccer team won 12 out of 14 games. If this rate continues, how many games will they win if they play a total of 21 games?

8. **VEGETABLES** At a harvest, 16 ears of corn are being picked for every 18 peppers. If 9 peppers have been picked, how many ears of corn have been picked?

9. **CONSTRUCTION** At a road work site, 20 cones are placed along 50 feet of road. At this rate, how many cones are placed along 35 feet of road?

Multi-Part Lesson 1

PART A

NAME ______________________________ DATE ____________ PERIOD _____

Reteach

Decimals as Fractions

Decimals like 0.58, 0.12, and 0.08 can be written as fractions.

To write a decimal as a fraction, you can follow these steps.

1. Identify the place value of the last decimal place.
2. Write the decimal as a fraction using the place value as the denominator, and simplify.

Example 1 **Write 0.5 as a fraction in simplest form.**

$0.5 = \frac{5}{10}$ 0.5 means five tenths.

$= \frac{\overset{1}{\cancel{5}}}{\underset{2}{\cancel{10}}}$ Simplify. Divide the numerator and denominator by the GCF, 5.

$= \frac{1}{2}$ So, in simplest form, 0.5 is $\frac{1}{2}$.

Example 2 **Write 0.35 as a fraction in simplest form.**

$0.35 = \frac{35}{100}$ 0.35 means 35 hundredths.

$= \frac{\overset{7}{\cancel{35}}}{\underset{20}{\cancel{100}}}$ Simplify. Divide the numerator and denominator by the GCF, 5.

$= \frac{7}{20}$ So, in simplest form, 0.35 is $\frac{7}{20}$.

Example 3 **Write 4.375 as a mixed number in simplest form.**

$4.375 = 4\frac{375}{1{,}000}$ 0.375 means 375 thousandths.

$= 4\frac{\overset{3}{\cancel{375}}}{\underset{8}{\cancel{1{,}000}}}$ Simplify. Divide the numerator and denominator by the GCF, 125.

$= 4\frac{3}{8}$ So, in simplest form, 4.375 is $4\frac{3}{8}$.

Exercises

Write each decimal as a fraction or mixed number in simplest form.

1. 0.9 **2.** 0.8 **3.** 0.27 **4.** 0.75

5. 0.34 **6.** 0.125 **7.** 0.035 **8.** 0.008

9. 1.4 **10.** 3.6 **11.** 6.28 **12.** 2.65

13. 12.05 **14.** 4.004 **15.** 23.205 **16.** 51.724

Multi-Part Lesson 1 PART A

NAME ________________ DATE ________ PERIOD ____

Skills Practice

Decimals as Fractions

Write each decimal as a fraction or mixed number in simplest form.

1. 0.6 **2.** 10.9 **3.** 0.08

4. 6.25 **5.** 4.125 **6.** 0.075

7. 9.35 **8.** 3.56 **9.** 8.016

10. 21.5 **11.** 0.055 **12.** 7.42

13. 5.006 **14.** 3.875 **15.** 1.29

16. 2.015 **17.** 6.48 **18.** 0.004

19. 4.95 **20.** 8.425 **21.** 9.74

22. 0.47 **23.** 5.019 **24.** 1.062

25. 3.96 **26.** 0.824 **27.** 20.8

28. 6.45 **29.** 4.672 **30.** 0.375

Multi-Part Lesson 1

PART B

NAME ______________________ DATE ____________ PERIOD _____

Reteach

Fractions as Decimals

Fractions with denominators that are factors of 10, 100, or 1,000 can be written as decimals using equivalent fractions. Any fraction can also be written as a decimal by dividing the numerator by the denominator.

Example 1 **Write $\frac{3}{5}$ as a decimal.**

Since 5 is a factor of 10, write an equivalent fraction with a denominator of 10.

$$\frac{3}{5} \overset{\times 2}{=} \frac{6}{10}$$

$$= 0.6$$

So, $\frac{3}{5} = 0.6$.

Example 2 **Write $\frac{3}{8}$ as a decimal.**

Divide.

```
    0.375
8 ) 3.000
   -2 4
     60
    -56
      40
     -40
       0
```

So, $\frac{3}{8} = 0.375$.

Exercises

Write each fraction or mixed number as a decimal.

1. $\frac{3}{10}$

2. $\frac{3}{4}$

3. $\frac{1}{4}$

4. $\frac{1}{5}$

5. $\frac{1}{8}$

6. $2\frac{1}{4}$

7. $\frac{6}{20}$

8. $\frac{9}{25}$

9. $1\frac{3}{8}$

10. $1\frac{5}{8}$

11. $3\frac{5}{16}$

12. $4\frac{9}{20}$

Multi-Part Lesson 1

PART B

NAME ______________________ DATE ____________ PERIOD _____

Skills Practice

Fractions as Decimals

Write each fraction or mixed number as a decimal.

1. $\frac{9}{10}$

2. $\frac{21}{100}$

3. $\frac{3}{4}$

4. $\frac{1}{2}$

5. $\frac{2}{5}$

6. $\frac{7}{10}$

7. $\frac{5}{8}$

8. $3\frac{7}{8}$

9. $9\frac{2}{5}$

10. $\frac{66}{200}$

11. $\frac{3}{20}$

12. $6\frac{5}{8}$

13. $5\frac{2}{5}$

14. $12\frac{3}{8}$

15. $10\frac{17}{20}$

16. $2\frac{7}{16}$

17. $3\frac{11}{16}$

18. $6\frac{4}{5}$

19. $1\frac{11}{25}$

20. $10\frac{1}{8}$

21. $2\frac{1}{16}$

22. $3\frac{19}{20}$

23. $5\frac{12}{75}$

24. $3\frac{24}{25}$

NAME ______________________ DATE ____________ PERIOD _____

Reteach

Percents and Decimals

To write a percent as a decimal, first rewrite the percent as a fraction with a denominator of 100. Then write the fraction as a decimal.

Example 1 **Write 23% as a decimal.**

$23\% = \frac{23}{100}$ Rewrite the percent as a fraction with a denominator of 100.

$= 0.23$ Write *23 hundredths* as a decimal.

Example 2 **Write 7% as a decimal.**

$7\% = \frac{7}{100}$ Rewrite the percent as a fraction with a denominator of 100.

$= 0.07$ Write *7 hundredths* as a decimal.

To write a decimal as a percent, first write the decimal as a fraction with a denominator of 100. Then write the fraction as a percent.

Example 3 **Write 0.44 as a percent.**

$0.44 = \frac{44}{100}$ Write *44 hundredths* as a fraction.

$= 44\%$ Write the fraction as a percent.

Example 4 **Write 0.65 as a percent.**

$0.65 = \frac{65}{100}$ Write *65 hundredths* as a fraction.

$= 65\%$ Write the fraction as a percent.

Exercises

Write each percent as a decimal.

1. 39% **2.** 57% **3.** 82%

4. 13% **5.** 8% **6.** 4%

Write each decimal as a percent.

7. 0.86 **8.** 0.36 **9.** 0.65

10. 0.2 **11.** 0.48 **12.** 0.17

NAME ______________________ DATE __________ PERIOD _____

Skills Practice

Percents and Decimals

Write each percent as a decimal.

1. 5%

2. 8%

3. 37%

4. 12%

5. 29%

6. 54%

7. 48%

8. 79%

9. 3%

10. 6%

11. 20%

12. 59%

13. 23%

14. 92%

15. 15%

16. 31%

Write each decimal as a percent.

17. 0.3

18. 0.7

19. 0.19

20. 0.74

21. 0.66

22. 0.52

23. 0.21

24. 0.81

25. 0.13

26. 0.36

27. 0.28

28. 0.45

29. 0.94

30. 0.34

31. 0.26

32. 0.99

NAME ________________________ DATE ____________ PERIOD _____

Reteach

Percents Greater Than 100% and Percents Less Than 1%

A percent greater than 100% equals a number greater than 1. A percent less than 1% equals a number less than 0.01 or $\frac{1}{100}$.

Examples **Write each percent as a decimal and as a mixed number or fraction in simplest form.**

1 **280%**

$280\% = \frac{280}{100}$ Definition of percent

$= 2.8 \text{ or } 2\frac{4}{5}$

2 **0.12%**

$0.12\% = \frac{0.12}{100}$ Definition of percent

$= 0.0012 \text{ or } \frac{3}{2{,}500}$

Examples **Write each decimal as a percent.**

3 **2.17**

$2.17 = 217.$ Multiply by 100.

$= 217\%$

4 **0.0034**

$0.0034 = 000.34$ Multiply by 100.

$= 0.34\%$

Exercises

Write each percent as a decimal and as a mixed number or fraction in simplest form.

1. 200% **2.** 750% **3.** 325%

4. 0.3% **5.** 0.8% **6.** 0.48%

Write each decimal as a percent.

7. 2.6 **8.** 19 **9.** 5.14

10. 0.008 **11.** 0.0014 **12.** 0.0067

NAME ______________________ DATE ____________ PERIOD _____

Skills Practice

Percents Greater Than 100% and Percents Less Than 1%

Write each percent as a decimal and as a mixed number or fraction in simplest form.

1. 900%

2. 150%

3. 675%

4. 245%

5. 120%

6. 0.2%

7. 0.08%

8. 0.12%

9. 0.35%

Write each decimal as a percent.

10. 3.9

11. 81

12. 25

13. 6.75

14. 2.81

15. 0.001

16. 0.0046

17. 0.0069

18. 0.0083

Write each number as a percent.

19. $6\frac{1}{2}$

20. $2\frac{1}{2}$

21. $5\frac{1}{4}$

22. $\frac{1}{200}$

23. $\frac{2}{250}$

24. $\frac{3}{500}$

NAME ______________________ DATE __________ PERIOD _____

Reteach

Compare and Order Fractions

To compare two fractions.

- Find the *least common denominator (LCD)* of the fractions; that is, find the least common multiple of the denominators.
- Write an equivalent fraction for each fraction using the LCD.
- Compare the numerators.

Example 1 **Replace ● with <, >, or = to make $\frac{1}{3}$ ● $\frac{5}{12}$ true.**

The LCM of 3 and 12 is 12. So, the LCD is 12.

Rewrite each fraction with a denominator of 12.

$\frac{1}{3} = \frac{\blacksquare}{12}$ (×4), so $\frac{1}{3} = \frac{4}{12}$. $\qquad \frac{5}{12} = \frac{5}{12}$

Now compare. Since $4 < 5$, $\frac{4}{12} < \frac{5}{12}$. So, $\frac{1}{3} < \frac{5}{12}$.

Example 2 **Order $\frac{1}{6}, \frac{2}{3}, \frac{1}{4}$, and $\frac{3}{8}$ from least to greatest.**

The LCD of the fractions is 24. So, rewrite each fraction with a denominator of 24.

$\frac{1}{6} = \frac{\blacksquare}{24}$ (×4), so $\frac{1}{6} = \frac{4}{24}$. $\qquad \frac{2}{3} = \frac{\blacksquare}{24}$ (×8), so $\frac{2}{3} = \frac{16}{24}$.

$\frac{1}{4} = \frac{\blacksquare}{24}$ (×6), so $\frac{1}{4} = \frac{6}{24}$. $\qquad \frac{3}{8} = \frac{\blacksquare}{24}$ (×3), so $\frac{3}{8} = \frac{9}{24}$.

The order of the fractions from least to greatest is $\frac{1}{6}, \frac{1}{4}, \frac{3}{8}, \frac{2}{3}$.

Exercises

Replace each ● with <, >, or = to make a true statement.

1. $\frac{5}{12}$ ● $\frac{3}{8}$
2. $\frac{6}{8}$ ● $\frac{3}{4}$
3. $3\frac{2}{7}$ ● $3\frac{1}{6}$

Order the fractions from least to greatest.

4. $\frac{3}{4}, \frac{3}{8}, \frac{1}{2}, \frac{1}{4}$
5. $\frac{2}{3}, \frac{1}{6}, \frac{5}{18}, \frac{7}{9}$
6. $1\frac{1}{2}, 1\frac{5}{6}, 1\frac{5}{8}, 1\frac{5}{12}$

NAME ______________________ DATE ____________ PERIOD ______

Multi-Part Lesson 3 PART B

Skills Practice

Compare and Order Fractions

Replace each ● with <, >, or = to make a true statement.

1. $\frac{2}{3}$ ● $\frac{3}{4}$

2. $\frac{3}{8}$ ● $\frac{6}{16}$

3. $\frac{5}{8}$ ● $\frac{7}{12}$

4. $\frac{1}{2}$ ● $\frac{6}{7}$

5. $\frac{3}{9}$ ● $\frac{1}{3}$

6. $\frac{1}{6}$ ● $\frac{9}{10}$

7. $\frac{5}{6}$ ● $\frac{7}{8}$

8. $\frac{5}{8}$ ● $\frac{5}{12}$

9. $\frac{4}{5}$ ● $\frac{2}{3}$

10. $\frac{6}{7}$ ● $\frac{4}{5}$

11. $\frac{5}{12}$ ● $\frac{3}{16}$

12. $\frac{3}{4}$ ● $\frac{2}{9}$

13. $\frac{5}{7}$ ● $\frac{7}{10}$

14. $\frac{2}{15}$ ● $\frac{1}{6}$

15. $\frac{5}{12}$ ● $\frac{2}{5}$

16. $2\frac{3}{10}$ ● $2\frac{5}{14}$

17. $5\frac{4}{9}$ ● $5\frac{3}{7}$

18. $1\frac{3}{5}$ ● $1\frac{5}{9}$

19. $4\frac{1}{6}$ ● $4\frac{2}{12}$

20. $1\frac{7}{9}$ ● $1\frac{4}{7}$

21. $8\frac{9}{10}$ ● $8\frac{11}{12}$

22. $3\frac{1}{4}$ ● $3\frac{2}{8}$

23. $6\frac{8}{9}$ ● $6\frac{7}{8}$

24. $11\frac{2}{9}$ ● $11\frac{4}{15}$

Order the fractions from least to greatest.

25. $\frac{3}{4}, \frac{2}{5}, \frac{5}{8}, \frac{1}{2}$

26. $\frac{1}{3}, \frac{2}{7}, \frac{3}{14}, \frac{1}{6}$

27. $\frac{2}{3}, \frac{4}{9}, \frac{5}{6}, \frac{7}{12}$

28. $\frac{4}{5}, \frac{2}{3}, \frac{13}{15}, \frac{7}{9}$

29. $8\frac{11}{12}, 8\frac{5}{6}, 8\frac{3}{4}, 8\frac{9}{16}$

30. $2\frac{7}{15}, 2\frac{3}{5}, 2\frac{5}{12}, 2\frac{1}{2}$

NAME _______________ DATE __________ PERIOD ____

Reteach

Compare and Order Fractions, Decimals, and Percents

Example 1 **Replace the ● with <, >, or = to make $\frac{4}{5}$ ● 0.9 a true statement.**

$\frac{4}{5}$ ● 0.9 — Write the statement.

0.8 ● 0.9 — Write $\frac{4}{5}$ as a decimal.

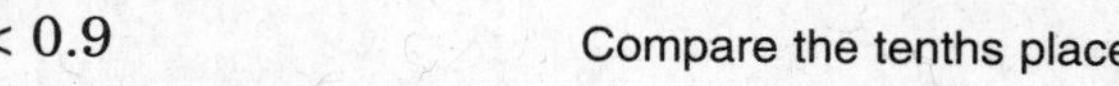

0.8 < 0.9 — Compare the tenths place.

Since 0.8 < 0.9, $\frac{4}{5}$ < 0.9.

Check Since 0.8 is to the left of 0.9 on the number line, $\frac{4}{5}$ < 0.9.

Example 2 **Replace the ● with <, >, or = to make 25% ● 0.23 a true statement.**

25% ● 0.23 — Write the statement.

0.25 ● 0.23 — Write 25% as a decimal.

0.25 > 0.23 — Since the tenths are the same, compare the hundredths.

Since 0.25 > 0.23, 25% > 0.23.

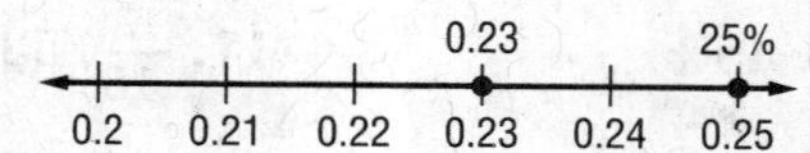

Check Use the number line.

Exercises

Replace the ● with <, >, or = to make a true statement.

1. $\frac{1}{2}$ ● 47%

2. 0.65 ● $\frac{13}{20}$

3. 89% ● $\frac{9}{10}$

4. $\frac{17}{19}$ ● 0.85

5. 34% ● $\frac{3}{10}$

6. 0.28 ● 25%

7. 0.75 ● 70%

8. 0.14 ● $\frac{1}{5}$

9. $\frac{3}{4}$ ● 85%

10. 0.3 ● $\frac{2}{5}$

NAME ____________ DATE ________ PERIOD ______

Skills Practice

Compare and Order Fractions, Decimals, and Percents

Replace the ● with <, >, or = to make a true statement.

1. $\frac{1}{2}$ ● 47%
2. 5.3 ● 530%
3. 35% ● $\frac{4}{7}$
4. $\frac{3}{5}$ ● 0.62
5. 80% ● 0.78
6. $\frac{2}{25}$ ● 6%
7. 0.53 ● $\frac{4}{9}$
8. 48% ● $\frac{5}{8}$
9. 93% ● 0.9

Order the numbers from least to greatest.

10. 0.44, $\frac{4}{7}$, 45%
11. $\frac{7}{15}$, 53%, 0.5
12. 0.66, $\frac{2}{3}$, 60%
13. $\frac{1}{4}$, 23%, 0.2
14. $\frac{5}{6}$, 80%, 0.9
15. $\frac{1}{5}$, 18%, 0.15

16. **FUNDRAISER** The table shows the sales performance for each grade level in the school fundraiser. Which grade raised the most money for the school?

Grade	Sales Performance
6	36%
7	$\frac{1}{4}$
8	0.39

NAME ______________________ DATE __________ PERIOD _____

Reteach

Estimate with Percents

The table below shows some commonly used percents and their fraction equivalents.

Percent-Fraction Equivalents				
$20\% = \frac{1}{5}$	$50\% = \frac{1}{2}$	$80\% = \frac{4}{5}$	$25\% = \frac{1}{4}$	$33\frac{1}{3}\% = \frac{1}{3}$
$30\% = \frac{3}{10}$	$60\% = \frac{3}{5}$	$90\% = \frac{9}{10}$	$75\% = \frac{3}{4}$	$66\frac{2}{3}\% = \frac{2}{3}$
$40\% = \frac{2}{5}$	$70\% = \frac{7}{10}$	$100\% = 1$		

Examples **Estimate each percent.**

 20% of 58

20% is $\frac{1}{5}$.

Round 58 to 60 since it is divisible by 5.

$\frac{1}{5}$ of 60 is 12.

So, 20% of 58 is about 12.

2 76% of 21.

76% is close to 75% or $\frac{3}{4}$.

Round 21 to 20 since it is divisible by 4.

$\frac{1}{4}$ of 20 is 5.

So, $\frac{3}{4}$ of 20 is 3×5 or 15.

Thus, 76% of 21 is about 15.

Example 3 **Isabel is reading a book that has 218 pages. She wants to complete 25% of the book by Friday. About how many pages should she read by Friday?**

25% is $\frac{1}{4}$. Round 218 to 220.

$\frac{1}{4}$ of 220 is 55.

So, Isabel should read about 55 pages by Friday.

Exercises

Estimate each percent.

1. 49% of 8

2. 24% of 27

3. 19% of 46

4. 62% of 20

5. 40% of 51

6. 81% of 32

7. **TIPS** Jodha wants to tip the pizza delivery person 20%. If the cost of the pizzas is \$15.99, what would be a reasonable amount to tip?

NAME ______________________ DATE ____________ PERIOD ______

Skills Practice

Estimate with Percents

Estimate each percent.

1. 50% of 39

2. 24% of 13

3. 19% of 31

4. 49% of 71

5. 27% of 81

6. 52% of 118

7. 19% of 94

8. 33% of 61

9. 58% of 5

10. 41% of 10

11. 75% of 17

12. 82% of 24

13. 73% of 61

14. 62% of 34

15. 38% of 42

16. 79% of 16

17. 91% of 82

18. 67% of 241

Estimate the percent of the figure that is shaded.

19.

20.

21.

NAME ______________________ DATE ____________ PERIOD _____

Reteach

Percent of a Number

Example 1 **Find 25% of 260.**

Method 1:

Write 25% as a fraction in simplest form.
Use the fraction in a multiplication problem.

$$25\% = \frac{25}{100} \text{ or } \frac{1}{4}$$

$$25\% \text{ of } 260 = \frac{1}{4} \times 260$$

$$= 65$$

So, 25% of 260 is 65.

Method 2:

Write 25% as a decimal.

Then write a multiplication problem.

$$25\% = 0.25$$

$$25\% \text{ of } 260 = 0.25 \times 260$$

$$= 65$$

Example 2 **Find 175% of 56.**

Method 1:

Write 175% as a fraction in simplest form.
Use the fraction in a multiplication problem.

$$175\% = \frac{175}{100} \text{ or } \frac{7}{4}$$

$$175\% \text{ of } 56 = \frac{7}{4} \times 56$$

$$= \frac{7}{\cancel{4}_1} \times \frac{\cancel{56}^{14}}{1}$$

$$= 98$$

So, 175% of 56 is 98.

Method 2:

Write 175% as a decimal.

Then write a multiplication problem.

$$175\% = 1.75$$

$$175\% \text{ of } 56 = 1.75 \times 56$$

$$= 98$$

Exercises

Find the percent of each number.

1. 48% of 50

2. 40% of 95

3. 75% of 116

4. 8% of 85

5. 98% of 30

6. 0.3% of 460

7. 15% of 342

8. 350% of 60

9. 0.25% of 500

10. 2.7% of 110

NAME ______________________ DATE ____________ PERIOD _____

Skills Practice

Percent of a Number

Find the percent of each number.

1. 15% of 82

2. 256% of 75

3. 0.5% of 50

4. 76% of 450

5. 85% of 30

6. 0.8% of 56

7. 16% of 75

8. 430% of 50

9. 0.44% of 375

10. 15% of 620

11. 250% of 24

12. 0.5% of 600

13. 65% of 140

14. 0.6% of 25

15. 0.75% of 50

16. 160% of 80

17. **YARDWORK** Mr. Simpson spent 5 hours working on his lawn. If he spent 25% of the time edging the flower beds, how much time did he spend edging?

18. **SHIPPING** Aurelia is buying a new computer. The shipping cost is 8% of the purchase price. If Aurelia's computer costs $585, how much will it cost to have it shipped?

NAME ______________________ DATE __________ PERIOD _____

Reteach

Problem-Solving Investigation: Solve a Simpler Problem

When solving problems, one strategy that is helpful is to *solve a simpler problem.* Using some of the information presented in the problem, you may be able to set up and solve a simpler problem.

You can use the *solve a simpler problem* strategy, along with the following four-step problem-solving plan, to solve a problem.

1 Understand – Read and get a general understanding of the problem.

2 Plan – Make a plan to solve the problem and estimate the solution.

3 Solve – Use your plan to solve the problem.

4 Check – Check the reasonableness of your solution.

Example

PUZZLES **Steven and Darshelle are putting together a 500-piece puzzle. So far they have 40% of the puzzle complete. How many pieces are left for them to fit into the puzzle?**

Understand We know the total number of pieces in the puzzle and that 40% of the pieces are already put together in the puzzle. We need to find the number of pieces left to fit into the puzzle.

Plan Solve a simpler problem to find 100% − 40% or 60% of the 500 pieces. First find 10% of 500 and then use the result to find 60% of 500.

Solve 10% or $\frac{1}{10}$ of 500 is 50.

So, 60% or $\frac{6}{10}$ of 500 is 6×50 or 300.

Steven and Darshelle still have 300 pieces left to fit in the puzzle.

Check We know that 40% or 4 out of every 10 pieces of the puzzle are already put together in the puzzle. Since $500 \div 10 \times 4 = 200$ pieces and $200 + 300 = 500$, the answer is correct.

Exercise

SCHOLARSHIPS Crosswood Elementary School received $450 in donations for its scholarship fund. If 30% of the contributions were from local businesses, how much money did local businesses contribute?

NAME ______________________ DATE ____________ PERIOD _____

Skills Practice

Problem-Solving Investigation: Solve a Simpler Problem

Solve. Use the *solve a simpler problem* strategy.

1. **SCHOOLS** A total of 350 students voted on whether a marlin or a panther should be the new school's mascot. If 30% of the students voted for the panther as the mascot, how many of the students voted for the panther?

2. **READING** Over the summer, Maggie plans to read one book the first week and double the number of books each week for the next 5 weeks. How many books will Maggie read in the sixth week?

3. **GEOGRAPHY** The total area of Michigan is 96,810 square miles. Of that, about 40% is water. About how many square miles of Michigan's area is land?

4. **ANIMALS** A spider travels at a speed of 1.17 miles per hour. At this rate, about how far can a spider travel in 3 hours?

NAME ______________________ DATE ____________ PERIOD _____

Reteach

Problem-Solving Investigation: Act It Out

When solving problems, one strategy that is helpful is to *act it out*. By using paper and pencil, a model, fraction strips, or any manipulative, you can often act out the situation. Then, by using your model, you can determine an answer to the situation.

You can use the *act it out* strategy, along with the following four-step problem-solving plan, to solve a problem.

1 Understand – Read and get a general understanding of the problem.

2 Plan – Make a plan to solve the problem and estimate the solution.

3 Solve – Use your plan to solve the problem.

4 Check – Check the reasonableness of your solution.

Example HOBBIES **This fall, Floyd is going to play one sport and take music lessons. He is deciding between playing football, cross country, or soccer. He is also deciding between guitar lessons or piano lessons. How many possible combinations are there of a sport and music lesson for Floyd?**

Understand You know the three sports he is choosing from: football, cross country, and soccer. You also know the music lessons he is choosing from: guitar and piano. You need to determine the number of possible combinations.

Plan Start by choosing one sport, and pairing it with each of the two music lessons. Then do this for each sport.

Solve

football, guitar	cross country, guitar	soccer, guitar
football, piano	cross country, piano	soccer, piano

So, there are 6 possible combinations of a sport and music lessons.

Check You can multiply the number of sport choices by the number of music lesson choices: $3 \times 2 = 6$.

Exercises

1. **WOODWORK** Tyson and his dad are making wooden picture frames. Each picture frame uses $1\frac{1}{4}$ feet of wood. If they have a total of $8\frac{1}{2}$ feet of wood, how many picture frames can they make?

2. **SEATING** Mom, Dad, and three children, Aja, Barry, and Cale, sat in the front row of a movie theater. The parents did not sit together. Barry did not sit next to Aja, and Cale did not sit next to his mother. Aja did not sit next to her dad. Show one way the family members could have arranged themselves.

NAME ______________________ DATE __________ PERIOD ______

Skills Practice

Problem-Solving Investigation: Act It Out

Solve. Use the *act it out* strategy.

1. **PAPER** A paper is folded in half four times. Once reopened, how many sections are there?

2. **JEWELRY** Brody is making a necklace, bracelet, and anklet out of beads. She has green, blue, purple, and silver beads. How many different pieces of jewelry can she make if she only uses one color of beads for each?

3. **CLOTHES** You can buy school uniforms through an online catalog. Boys can order either navy blue or khaki pants with a red, white, or blue shirt. How many uniform combinations are there online for boys?

4. **TIME** School is out at 3:45 P.M., band practice is $2\frac{1}{2}$ hours, dinner takes 45 minutes, and you go to bed at 10:00 P.M. How much free time will you have if you study for 2 hours for a math exam?

NAME ______________________ DATE ____________ PERIOD _____

Reteach

Algebra: Properties

Property	Symbols	Numbers
Commutative	$a + b = b + a$ $a \cdot b = b \cdot a$	$5 + 3 = 3 + 5$ $5 \cdot 3 = 3 \cdot 5$
Associative	$(a + b) + c = a + (b + c)$ $(a \cdot b) \cdot c = a \cdot (b \cdot c)$	$(2 + 3) + 4 = 2 + (3 + 4)$ $(2 \cdot 3) \cdot 4 = 2 \cdot (3 \cdot 4)$
Identity	$a + 0 = a$ $a \cdot 1 = a$	$5 + 0 = 5$ $5 \cdot 1 = 5$

Example 1 **Determine whether 6 + (4 + 3) and (6 + 4) + 3 are equivalent.**

The numbers are grouped differently. They are equivalent by the Associative Property. So, $6 + (4 + 3) = (6 + 4) + 3$.

Use the properties to make mental math easier.

Example 2 **The formula for the perimeter of a triangle is $P = a + b + c$, where a, b, and c are side lengths. Find the perimeter of a triangle where $a = 12$, $b = 5$, and $c = 8$.**

$P = a + b + c$	Write the formula.
$P = 12 + 5 + 8$	Replace *a* with 12, *b* with 5, and *c* with 8.
$P = 12 + 8 + 5$	Commutative Property
$P = 25$ units	

Exercises

Determine whether the two expressions are equivalent. If so, tell what property is applied. If not, explain why.

1. $9 \cdot 1$ and 9

2. $7 \cdot 3$ and $3 \cdot 7$

3. $6 - (3 - 2)$ and $(6 - 3) - 2$

4. $(10 \cdot 2) \cdot 5$ and $10 \cdot (2 \cdot 5)$

5. The formula for the volume of a rectangular prism is $V = \ell wh$, where ℓ is the length, w is the width, and h is the height. Find the volume of a rectangular prism, in cubic units, if the length is 15 units, the width is 8 units, and the height is 2 units. Use the Commutative Property.

Multi-Part Lesson 2 PART A

NAME ______________________ DATE ____________ PERIOD ______

Skills Practice

Algebra: Properties

Determine whether the two expressions are equivalent. If so, tell what property is applied. If not, explain why.

1. $2 \cdot (3 \cdot 7)$ and $(2 \cdot 3) \cdot 7$

2. $6 + 3$ and $3 + 6$

3. $26 - (9 - 7)$ and $(26 - 9) - 7$

4. $18 \cdot 1$ and 18

5. $7 \cdot 2$ and $2 \cdot 7$

6. $6 - (4 - 1)$ and $(6 - 4) - 1$

7. $7 + 0$ and 7

8. $0 + 12$ and 0

9. $625 + 281$ and $281 + 625$

10. $(12 \cdot 18) \cdot 5$ and $12 \cdot (18 \cdot 5)$

11. $2 + (8 + 2)$ and $(2 + 8) + 2$

12. $40 \div 10$ and $10 \div 40$

Use one or more properties to rewrite each expression as an expression that does not use parentheses.

13. $(p \cdot 1) \cdot 6$

14. $(a + 5) + 23$

15. $7 \cdot (y \cdot 3)$

16. $(b + 4) + 17$

17. $6 + (x + 50)$

18. $(y \cdot 200) \cdot 2$

NAME ______________________ DATE __________ PERIOD _____

Multi-Part Lesson 2 PART C

Reteach

The Distributive Property

- To multiply a sum by a number, multiply each addend by the number outside the parentheses.
- $a(b + c) = ab + ac$
- $(b + c)a = ba + ca$

Example 1 **Find 6 × 38 mentally using the Distributive Property.**

$6 \times 38 = 6(30 + 8)$	Write 38 as 30 + 8.
$= 6(30) + 6(8)$	Distributive Property
$= 180 + 48$	Multiply mentally.
$= 228$	Add.

So, $6 \times 38 = 228$.

Example 2 **Use the Distributive Property to rewrite $4(x + 3)$.**

$4(x + 3) = 4(x) + 4(3)$	Distributive Property
$= 4x + 12$	Multiply.

So, $4(x + 3)$ can be rewritten as $4x + 12$.

Exercises

Find each product mentally. Show the steps you used.

1. 4×82

2. 9×26

3. 12×44

4. 8×5.7

Use the Distributive Property to rewrite each algebraic expression.

5. $5(y + 4)$

6. $(7 + r)3$

7. $12(x + 5)$

8. $(b + 2)9$

9. $4(4 + a)$

10. $9(7 + v)$

NAME ______________________ DATE __________ PERIOD _____

Skills Practice

The Distributive Property

Find each product mentally. Show the steps you used.

1. 3×78

2. 7×74

3. 8×92

4. 6×57

5. $15 \times 2\frac{2}{3}$

6. $12 \times 5\frac{1}{6}$

7. 6×5.2

8. 4×9.4

Use the Distributive Property to rewrite each algebraic expression.

9. $7(y + 2)$

10. $(8 + r)4$

11. $8(x + 9)$

12. $(b + 5)12$

13. $4(2 + a)$

14. $7(6 + v)$

15. $(b + 5)15$

16. $3(5 + v)$

17. $6(11 + s)$

Multi-Part Lesson 1 PART A

NAME ______________________ DATE ____________ PERIOD _____

Reteach

Equations

An **equation** is a mathematical sentence showing two expressions are equal. An equation contains an **equals sign**, =. Some equations contain variables. When you replace a variable with a value that results in a true sentence, you **solve** the equation. The value for the variable is the **solution** of the equation.

Example 1 **Solve $14 - p = 6$ using guess, check, and revise.**

Guess the value of p, then check it.

Try 7.	Try 6.	Try 8.
$14 - p = 6$	$14 - p = 6$	$14 - p = 6$
$14 - 7 \neq 6$	$14 - 6 \neq 6$	$14 - 8 = 6$
revise	revise	yes

The solution is 8 because replacing p with 8 results in a true sentence.

Example 2 **Solve $15 \div m = 5$ mentally.**

$15 \div m = 5$ — **Think** 15 divided by what number is 5?

$15 \div 3 = 5$ — You know that $15 \div 3 = 5$.

$5 = 5$

The solution is 3.

Exercises

Identify the solution of each equation from the list given.

1. $h + 19 = 56$; 36, 37, 38
2. $31 + x = 42$; 9, 10, 11
3. $k - 4 = 13$; 16, 17, 18
4. $34 - b = 17$; 16, 17, 18
5. $5w = 30$; 5, 6, 7
6. $63 = 7k$; 7, 8, 9
7. $36 \div s = 9$; 4, 5, 6
8. $x \div 3 = 8$; 23, 24, 25

Solve each equation mentally.

9. $j + 3 = 9$
10. $14 + n = 19$
11. $23 + x = 29$
12. $31 - h = 24$
13. $m - 5 = 11$
14. $3m = 27$
15. $56 = 7b$
16. $14 \div f = 2$
17. $j \div 8 = 4$

Multi-Part Lesson 1 PART A

NAME ______________________ DATE ____________ PERIOD ______

Skills Practice

Equations

Identify the solution of each equation from the list given.

1. $s + 12 = 17$; 5, 6, 7

2. $54 + f = 70$; 16, 17, 18

3. $69 = 50 + s$; 17, 18, 19

4. $47 = 77 - b$; 20, 30, 40

5. $44 = t - 10$; 52, 53, 54

6. $25 - k = 20$; 5, 6, 7

7. $4r = 40$; 8, 9, 10

8. $33 = 11d$; 3, 4, 5

9. $6g = 36$; 5, 6, 7

10. $28 \div w = 7$; 3, 4, 5

11. $b \div 6 = 4$; 22, 23, 24

12. $56 \div c = 8$; 6, 7, 8

Solve each equation mentally.

13. $4 + k = 11$

14. $7 + f = 15$

15. $z + 16 = 25$

16. $j + 15 = 30$

17. $20 = 30 - n$

18. $34 = r - 10$

19. $23 - m = 10$

20. $p - 4 = 13$

21. $8w = 80$

22. $7q = 21$

23. $48 = 6g$

24. $54 = 9m$

25. $18 \div t = 6$

26. $y \div 3 = 5$

27. $h \div 12 = 1$

28. **FLOWERS** Mickaela picked flowers for her shop. She picked 12 dozen flowers in the morning. By the end of the day, she had picked 18 dozen flowers. Use mental math or the *guess, check, and revise* strategy to solve the equation $12 + d = 18$, and find d, the number of dozens of flowers picked during the rest of the day.

NAME ______________________ DATE ____________ PERIOD _____

Reteach

Problem-Solving Investigation: Work Backward

When solving problems, one strategy that is helpful is to *work backward.* Sometimes you can use information in the problem to work backward to find what you are looking for, or the answer to the problem.

You can use the *work backward* strategy, along with the following four-step problem-solving plan to solve a problem.

1 Understand – Read and get a general understanding of the problem.

2 Plan – Make a plan to solve the problem and estimate the solution.

3 Solve – Use your plan to solve the problem.

4 Check – Check the reasonableness of your solution.

Example TIME **Meagan is meeting her friends at the library at 6:30 P.M. Before her mom takes her to the library, they are going to stop by her grandma's house. It takes 15 minutes to get from her house to her grandma's house, and they will stay and visit for 30 minutes. If it takes 5 minutes to get from her grandma's house to the library, what time should Meagan and her mom leave their house?**

Understand We know the time Meagan is meeting her friends at the library. We need to find what time Meagan and her mom should leave their house.

Plan To find the time they should leave, start with the 6:30 P.M. and first subtract 5 minutes for the time it takes to get from her grandma's house to the library.

Solve

Time from grandma's to library: 6:30 P.M. – 5 minutes = 6:25 P.M.
Time visiting with grandma: 6:25 P.M. – 30 minutes = 5:55 P.M.
Time from home to grandma's: 5:55 P.M. – 15 minutes = 5:40 P.M.

Meagan and her mom should leave their house at 5:40 P.M.

Check Add up all the times: 15 min + 30 min + 5 min = 50 min. When you add 50 minutes to 5:40 P.M., the result is 6:30 P.M., so the answer is correct.

Exercises

Solve using the *work backward* strategy.

1. NUMBER SENSE A number is divided by 3. Next, 7 is added to the quotient. Then, 10 is subtracted from the sum. If the result is 5, what is the number?

2. FAIR Timur paid $4.50 to enter the fair. He then bought a pretzel for $1.50 and played a game for $1. Last, he rode a roller coaster for $3. If he came home with $8, how much did he take to the fair?

NAME ______________________________ DATE ____________ PERIOD _____

Multi-Part Lesson 1

PART B

Skills Practice

Problem-Solving Investigation: Work Backward

Solve using the *work backward* strategy.

1. **MONEY** Lindsay bought 2 pairs of shoes that were the same price. Including the $3 sales tax, she paid a total of $57. What was the cost of each pair of shoes before the tax was added?

2. **TIME** Hung has to be at school by 7:10 A.M. It takes 20 minutes for him to shower and get dressed, and 15 minutes to eat breakfast. If Hung has a 25-minute bus ride to school, what is the latest time he should get up in the morning?

3. **NUMBER SENSE** A number is multiplied by 4. Then 7 is added to the product. After subtracting 3, the result is 8. What is the number?

4. **SCIENCE** A certain bacteria doubles its population every 12 hours. After 3 days, there were 1,600 bacteria. How many bacteria were there at the beginning of the first day?

Multi-Part Lesson 3 PART B

NAME ______________________ DATE ____________ PERIOD ______

Reteach

Solve and Write Two-Step Equations

A **two-step equation** is an equation with two different operations. To solve a two-step equation, undo the operations in reverse order of the order of operations. Therefore, undo the addition or subtraction first. Then undo the multiplication or division.

Example 1 **Solve $3x + 5 = 8$.**

$3x + 5 = 8$	Write the equation.
$3x + 5 = 8$	Undo the addition first by subtracting 5 from each side.
$\underline{\quad -5 \;\; -5}$	
$3x \quad = 3$	
$\frac{3x}{3} = \frac{3}{3}$	Divide each side by 3.
$x = 1$	Simplify.

Check:

$3x + 5 = 8$	Write the original equation.
$3(1) + 5 \stackrel{?}{=} 8$	Replace *x* with 1.
$3 + 5 \stackrel{?}{=} 8$	Simplify.
$8 = 8$	The sentence is true. ✓

Example 2 **Jules found $3 worth of coins in her couch. This is $1 less than twice the amount she found in her car. How much money did she find in her car?**

Words Twice the amount in the car less $1 is $3.

Variable Let c = amount in car.

Equation $2c - 1 = 3$

$2c - 1 = 3$	Write the equation.
$\underline{\quad +1 \;\; +1}$	Add 1 to each side.
$2c \quad = 4$	
$\frac{2c}{2} = \frac{4}{2}$	Divide each side by 2.
$c = 2$	Simplify.

Jules found $2 worth of coins in her car.

Exercises

Solve each equation. Check your solution.

1. $2a + 10 = 24$
2. $3c - 6 = 12$
3. $5 + 2g = 11$
4. $4y + 1 = 25$
5. $2t - 21 = 11$
6. $10 + 5d = 75$

NAME ______________________ DATE ____________ PERIOD ______

Multi-Part Lesson 3 PART B

Skills Practice

Solve and Write Two-Step Equations

Solve each equation. Check your solution.

1. $5x - 4 = 11$

2. $6b + 3 = 21$

3. $4n - 7 = 9$

4. $3c + 4 = 34$

5. $8v - 10 = 70$

6. $7j + 6 = 62$

7. $4v - 8 = 44$

8. $20 + 5p = 75$

9. $6z - 8 = 88$

10. $4q + 4 = 40$

11. $5d - 15 = 65$

12. $4 + 7u = 60$

13. $9v - 8 = 91$

14. $19 + 6p = 79$

15. $11j - 12 = 120$

16. $15 + 5p = 100$

17. $12q - 6 = 90$

18. $22 + 10p = 92$

19. WORKING Cameron made $50 working at a golf course. He made $10 cleaning golf balls, and then he worked as a caddy for 2 hours. The equation $10 + 2c = 50$ can be used to find how much Cameron made caddying per hour. Find this amount.

20. BATTERIES Research shows that a new battery will last for 36 hours. This is 2 more hours than twice the amount of time that the old battery lasted. Solve the equation $2h + 2 = 36$ to find how long the old battery lasted.

21. MOTORCYCLES Mr. Mendez bought 2 tickets to see the motorcross games with his son. He used the $5 coupon he found in his paper. He spent $25 for the tickets. The equation $2t - 5 = 25$ can be used to find the cost of one ticket. Find the cost of one ticket.

Multi-Part Lesson 1 PART A

NAME ______________________ DATE ____________ PERIOD ______

Reteach

Graph Relations

A **coor**[illegible] **plane** is formed when two number lines intersect at their zero points. This intersection is called t[illegible]. The horizontal number line is called the **x-axis**. The vertical number line is called the ***y*-axis**.

An **ordered pair** i[illegible] name a point on a coordinate plane. The first number in the ordered pair is the **x-coordinate**, an[illegible]cond number is the ***y*-coordinate**.

Example 1 **Graph the** [illegible] ***W*(2, 4).**

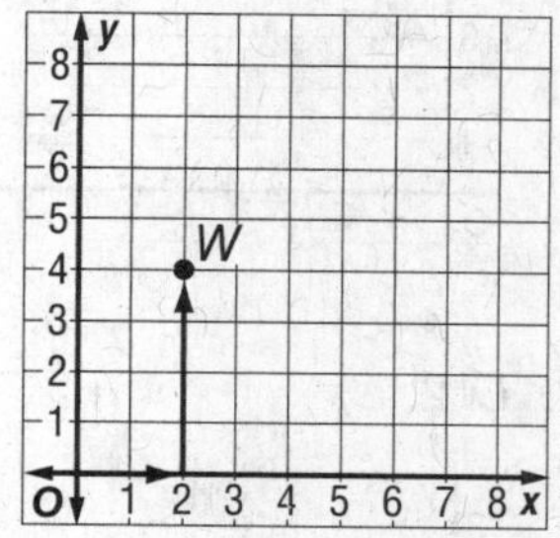

Start at the origin. Move 2 units [illegible] right along the x-axis.

Then move 4 units up to locate the p[illegible]aw a dot and label the point W.

Example 2 TICKETS **Tickets to the school play c**[illegible] **each. The costs of 1, 2, and 3 tickets are sho**[illegible] **the table. List this information as ordered pa**[illegible] **(number of tickets, cost).**

The ordered pairs are (1, 3), (2, 6), and (3, 9).

Ticket Costs	
Number of Tickets	Cost ($)
1	3
2	6
[illegible]	9

Example 3 **Graph the ordered pairs from Example 2.**

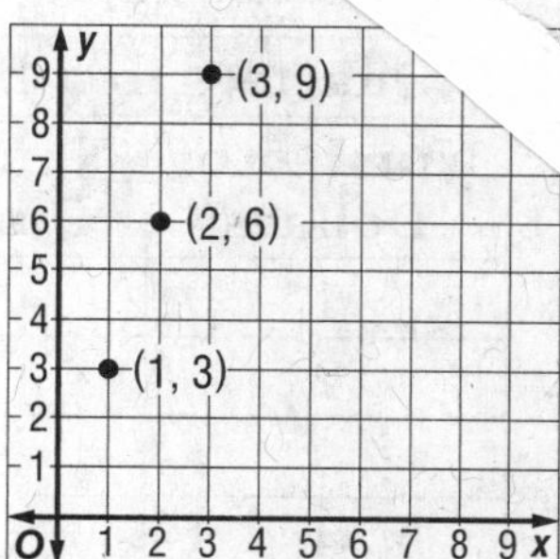

Exercises

Graph and label each point on the coordinate plane.

1. $S(1, 3)$

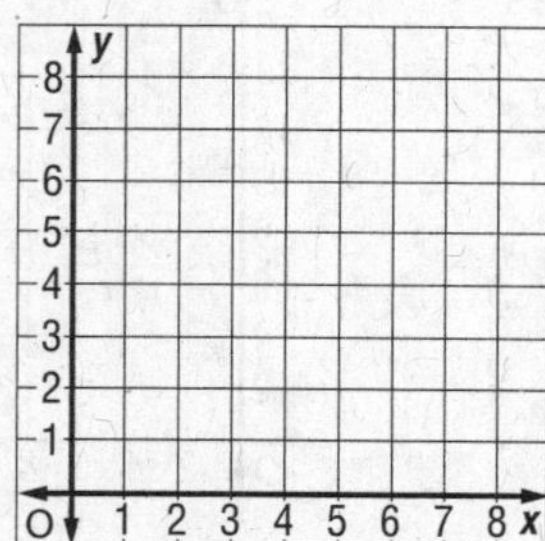

2. $T(4, 0)$

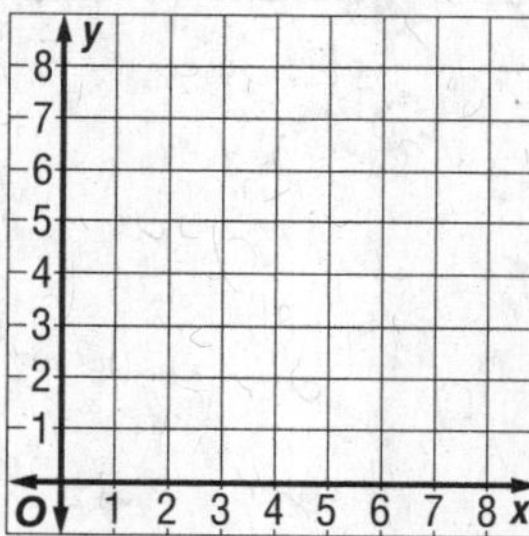

NAME ______________________ DATE ____________ PERIOD ______

Skills Practice

Graph Relations

Graph and label each point on the coordinate plane.

1. $A(1, 3)$

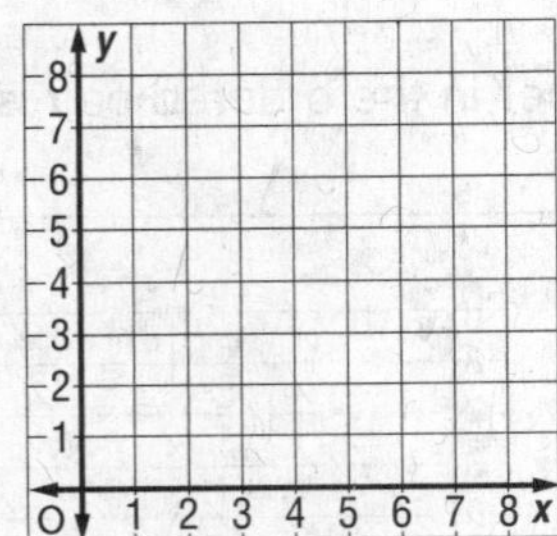

2. $B(4, 3)$

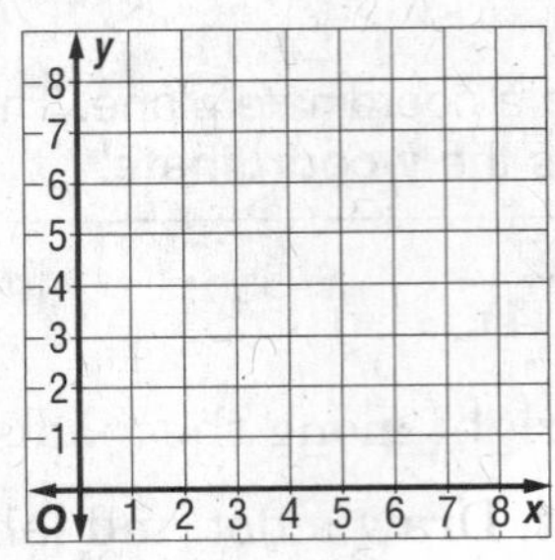

3. $C(2, 0)$

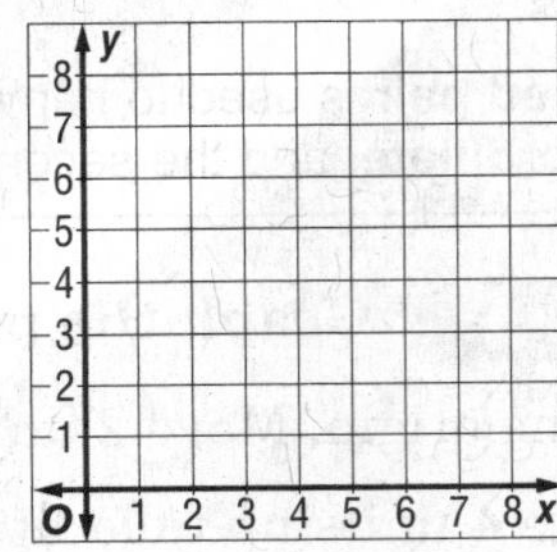

4. $D(2, 5)$

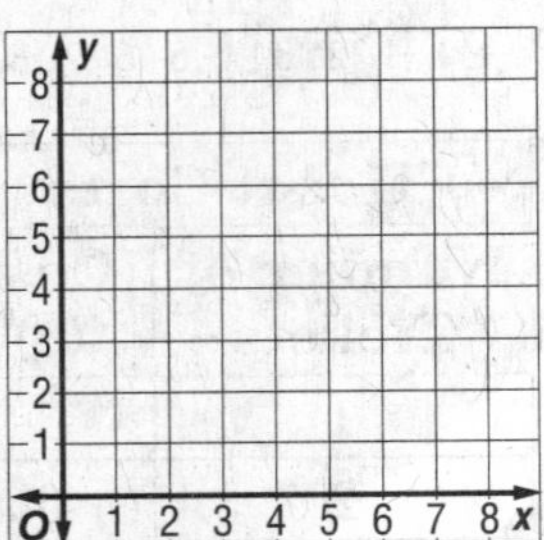

5. $E(5, 3)$

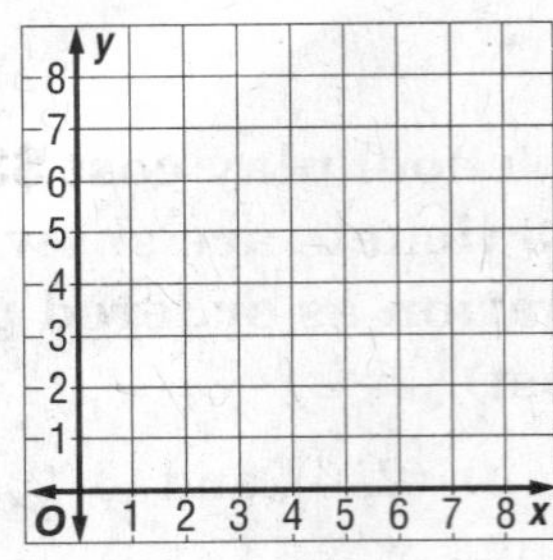

6. $F(3, 4)$

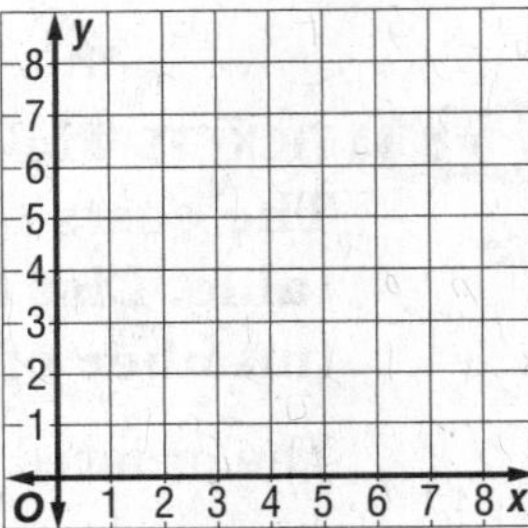

7. **MONEY** One dollar is worth 4 quarters. The table below shows this relationship.

Quarters in a Dollar	
Number of Dollars	Number of Quarters
1	4
2	8
3	12
4	16

a. List this information as ordered pairs (number of dollars, number of quarters).

b. Graph the ordered pairs. Then describe the graph.

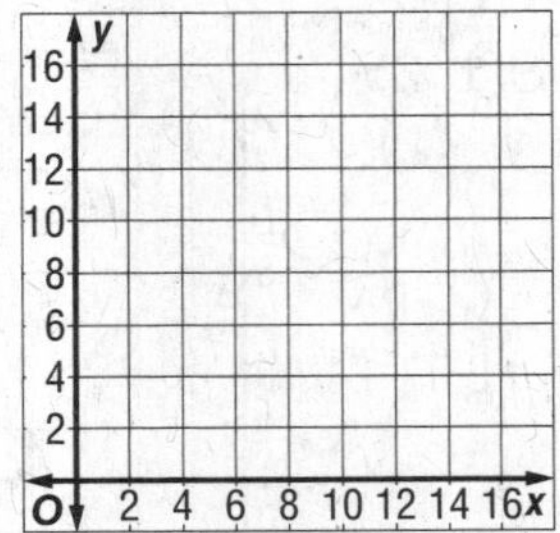

NAME ______________________ DATE ____________ PERIOD _____

Reteach

Function Tables

A **function rule** describes the relationship between the input and output of a **function**. The inputs and outputs can be organized in a **function table**.

Example 1 **Complete the function table.**

Input (x)	$x - 3$	Output (y)
9	$9 - 3$	■
8	$8 - 3$	■
6	$6 - 3$	■

The function rule is $x - 3$. Subtract 3 from each input.

Input **Output**

9 $-3 \rightarrow$ 6

8 $-3 \rightarrow$ 5

6 $-3 \rightarrow$ 3

$\rightarrow$

Input (x)	$x - 3$	Output (y)
9	$9 - 3$	6
8	$8 - 3$	5
6	$6 - 3$	3

Example 2 **Find the input for the function table.**

Input (x)	$4x$	Output (y)
■	■	0
■	■	4
■	■	8

Work backward to find the input. Since the rule is $4x$, divide each output by 4.

The inputs are 0, 1, and 2.

Exercises

Complete each function table.

1.

Input (x)	$2x$	Output (y)
0		
2		
4		

2.

Input (x)	$4 + x$	Output (y)
0		
1		
4		

Find the input for each function table.

3.

Input (x)	$x + 2$	Output (y)
	$1 + 2$	3
	$2 + 2$	4
	$5 + 2$	7

4.

Input (x)	$x \div 2$	Output (y)
	$2 \div 2$	1
	$6 \div 2$	3
	$10 \div 2$	5

NAME ______________________ DATE __________ PERIOD ____

Skills Practice

Function Tables

Complete each function table.

1.

Input (x)	$x + 3$	Output (y)
0		
2		
4		

2.

Input (x)	$3x + 1$	Output (y)
0		
1		
2		

3.

Input (x)	$2x - 1$	Output (y)
7		
5		
4		

4.

Input (x)	$x \div 3$	Output (y)
12		
9		
6		

5. If a function rule is $2x - 3$, what is the output for the input 3?

6. If a function rule is $4 - x$, what is the output for the input 2?

Find the input for each function table.

7.

Input (x)	$x - 3$	Output (y)
		7
		4
		1

8.

Input (x)	$x + 9$	Output (y)
		12
		15
		17

9.

Input (x)	$5x$	Output (y)
		0
		10
		15

10.

Input (x)	$x \div 2$	Output (y)
		2
		3
		6

11.

Input (x)	$2x + 2$	Output (y)
		4
		6
		8

12.

Input (x)	$3x - 1$	Output (y)
		14
		8
		5

Multi-Part Lesson 1 PART D

NAME ______________________ DATE ____________ PERIOD ____

Reteach

Function Rules

A **sequence** is a list of numbers in a specific order. Each number in the sequence is called a **term**. An **arithmetic sequence** is a sequence in which each term is found by adding the same number to the previous term.

Example **Use words and symbols to describe the value of each term as a function of its position. Then find the value of the tenth term in the sequence.**

Position	1	2	3	4	n
Value of Term	4	8	12	16	■

Study the relationship between each position and the value of its term.

Notice that the value of each term is 4 times its position number. So, the value of the term in position n is $4n$.

To find the value of the tenth term, replace n with 10 in the algebraic expression $4n$. Since $4 \times 10 = 40$, the value of the tenth term in the sequence is 40.

Position		Value of term
1	$\times 4 =$	4
2	$\times 4 =$	8
3	$\times 4 =$	12
4	$\times 4 =$	16
n	$\times 4 =$	$4n$

Exercises

Use words and symbols to describe the value of each term as a function of its position. Then find the value of the tenth term in the sequence.

1.

Position	3	4	5	6	n
Value of Term	1	2	3	4	■

2.

Position	1	2	3	4	n
Value of Term	5	10	15	20	■

3.

Position	4	5	6	7	n
Value of Term	11	12	13	14	■

Multi-Part Lesson 1
PART D

NAME ______________________ DATE ____________ PERIOD _____

Skills Practice

Function Rules

Use words and symbols to describe the value of each term as a function of its position. Then find the value of the tenth term in the sequence.

1.

Position	5	6	7	8	n
Value of Term	2	3	4	5	■

2.

Position	1	2	3	4	n
Value of Term	6	12	18	24	■

3.

Position	1	2	3	4	n
Value of Term	10	11	12	13	■

4.

Position	1	2	3	4	n
Value of Term	5	10	15	20	■

Find a rule for each function table.

5.

Input (x)	Output (y)
5	0
6	1
7	2
8	3
x	■

6.

Input (x)	Output (y)
2	14
4	16
6	18
8	20
x	■

7.

Input (x)	Output (y)
4	0
5	1
6	2
7	3
x	■

8.

Input (x)	Output (y)
1	11
2	22
3	33
4	44
x	■

Multi-Part Lesson 1 PART E

NAME ______________________ DATE __________ PERIOD ____

Reteach

Functions and Equations

A **function table** displays **input** and **output** values that represent a function. The function displayed in a function table can be represented with an **equation**.

Example 1 **Write an equation to represent the function.**

Examine how the value of each input and output changes.

Input, x	1	2	3	4	5
Output, y	5	10	15	20	25

Each output y is equal to 5 times the input x.

So, the equation that represents the function is $y = 5x$.

	+1	+1	+1	+1	
Input, x	1	2	3	4	5
Output, y	5	10	15	20	25
	+5	+5	+5	+5	

Example 2 **Graph the equation $y = 5x$.**

Select any three values for the input x, for example, 0, 1, and 2. Substitute these values for x to find the output y.

x	$5x$	y	(x, y)
0	5(0)	0	(0, 0)
1	5(1)	5	(1, 5)
2	5(2)	10	(2, 10)

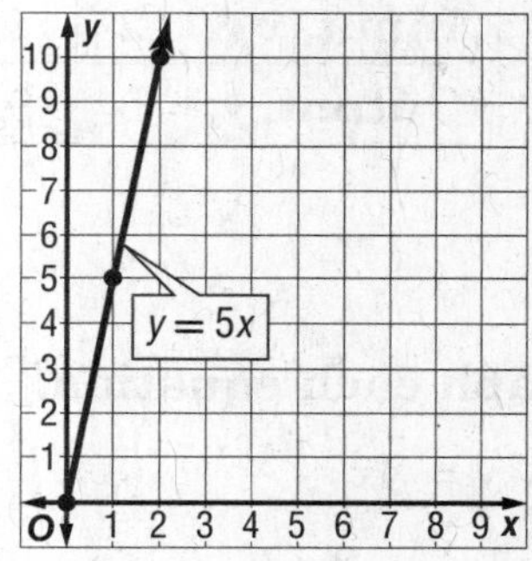

The ordered pairs (0, 0), (1, 5), and (2, 10) represent the function. They are solutions of the equation.

Exercises

Write an equation to represent each function.

1.

Input, x	1	2	3	4
Output, y	2	4	6	8

2.

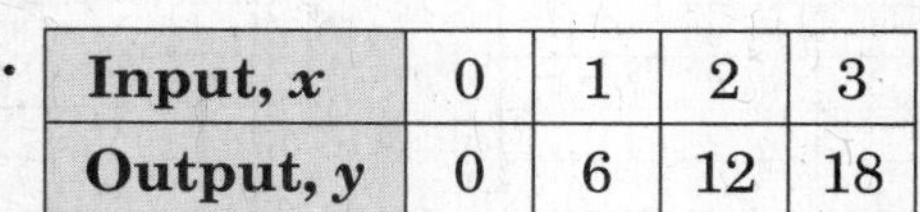

Input, x	0	1	2	3
Output, y	0	6	12	18

In Exercise 3, graph the equation you got as an answer in Exercise 1.
In Exercise 4, graph the equation you got in Exercise 2.

3. 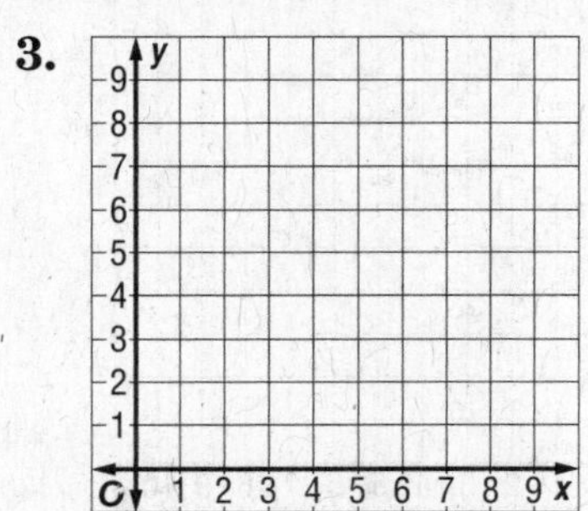

4. 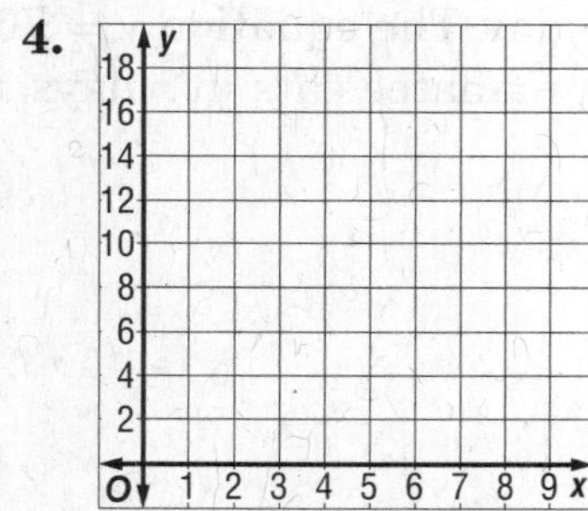

NAME ____________________ DATE __________ PERIOD _____

Skills Practice

Functions and Equations

Write an equation to represent each function.

1.

Input, x	0	1	2	3	4
Output, y	0	3	6	9	12

2.

Input, x	0	1	2	3	4
Output, y	0	1	2	3	4

3.

Input, x	1	2	3	4	5
Output, y	7	14	21	28	35

4.

Input, x	0	1	2	3	4
Output, y	7	8	9	10	11

5.

Input, x	2	4	6	8	10
Output, y	5	9	13	17	21

6.

Input, x	0	1	2	3	4
Output, y	2	14	26	38	50

Graph each equation.

7. $y = 4x$

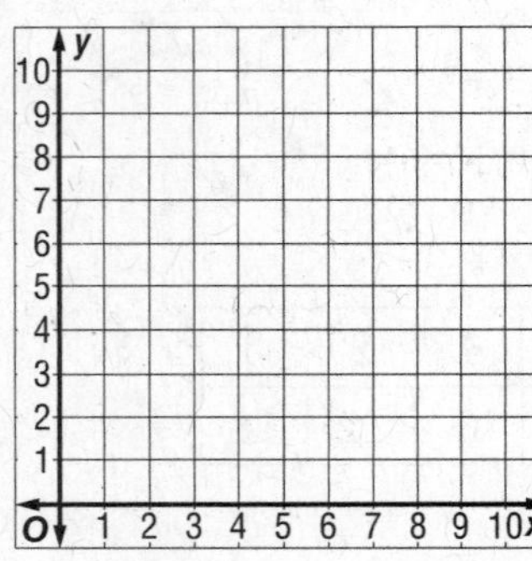

8. $y = x + 6$

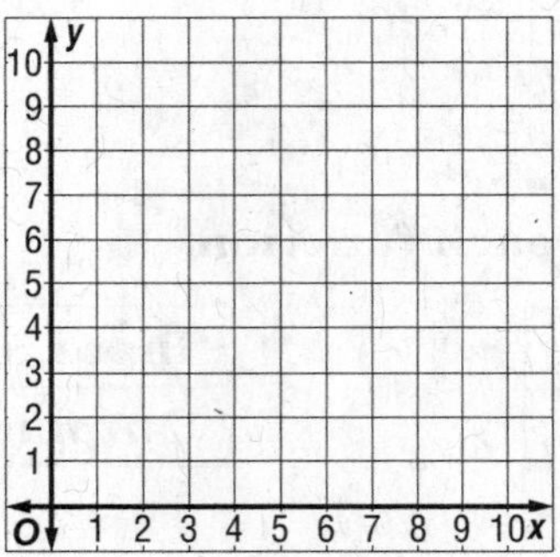

9. $y = 2x + 1$

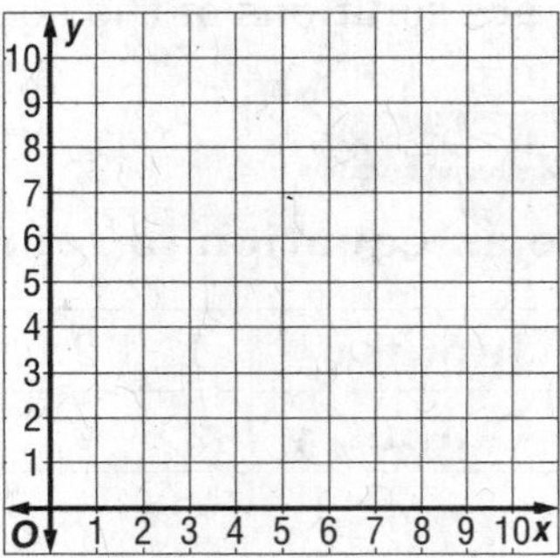

10. **ANIMALS** A manatee eats an average of 70 pounds of wet vegetation each day. The equation $y = 70x$ describes the amount y that a manatee eats in x days. Graph the function.

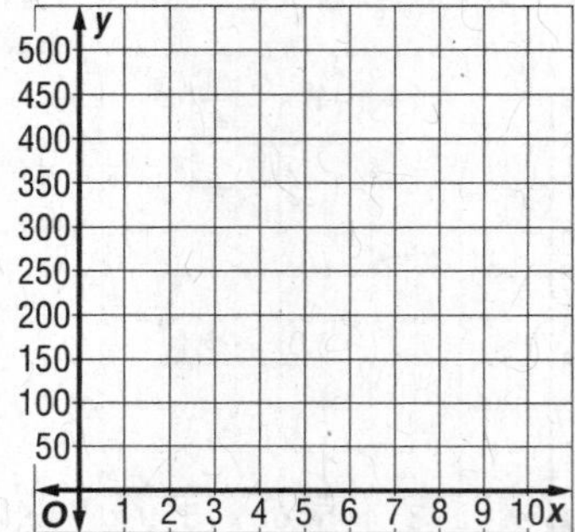

NAME ______________________ DATE ____________ PERIOD _____

Reteach

Multiple Representations of Functions

	Multiple Representations of Functions
Words The cost of movie tickets is equal to \$8 times the number of tickets.	**Equation** $c = 8t$

Table

Number of Tickets	Cost (\$)
1	8
2	16
3	24

Graph

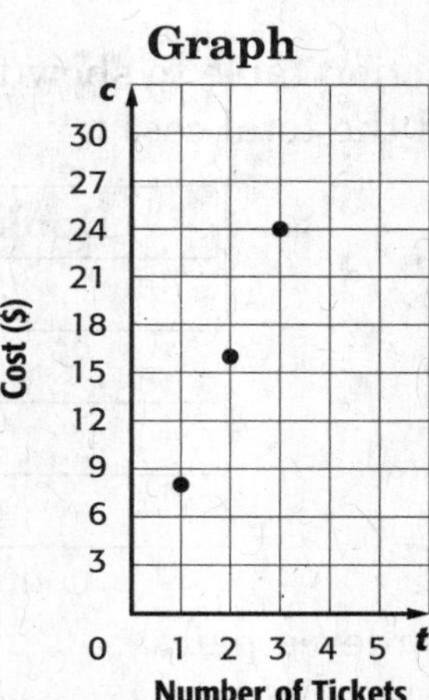

Exercises

1. **MINI GOLF** Great Big Fun charges \$3 per game of miniature golf.

   a. Write an equation to represent the total cost t of any number of games g of miniature golf.

   b. Make a function table to show the relationship between the number of games g and the total cost t.

Number of Games (g)	Total Cost (t)
1	
2	
3	
4	

   c. Graph the ordered pairs.

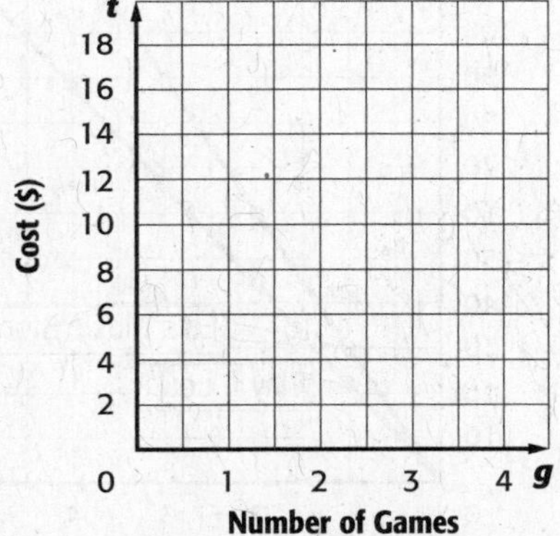

NAME ______________________ DATE ____________ PERIOD ______

Skills Practice

Multiple Representations of Functions

BOOKS The school library is buying new books that cost $9 each.

1. Write an equation to find t, the total cost in dollars for any number of books b.

2. Make a function table to show the relationship between the number of books b and the total cost t.

Books (b)	Total Cost (t)

3. Graph the ordered pairs.

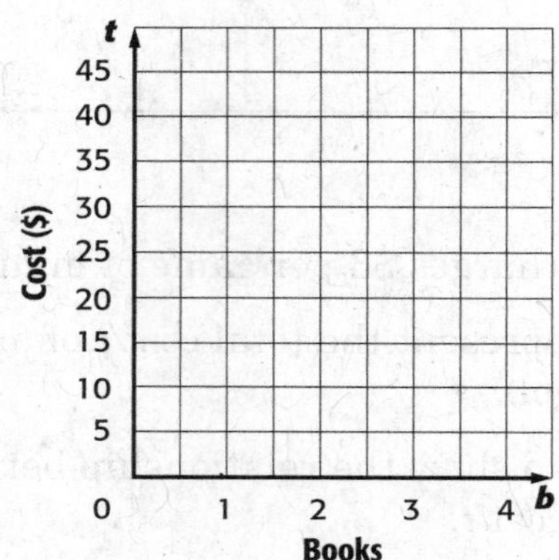

PIANO Play It Loud charges $22 per lesson for piano lessons. Ed's Music Store charges $30 per lesson.

4. Write an equation to represent the total cost t of any number of piano lessons p at each place.

5. Use the equations of the functions to determine which line is steeper. Explain your reasoning.

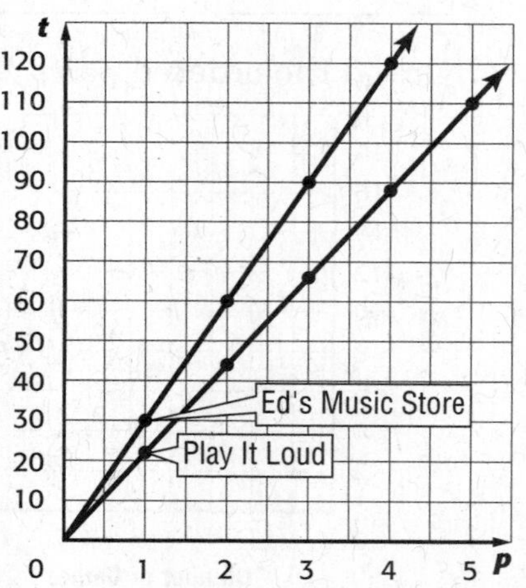

NAME ______________________ DATE __________ PERIOD _____

Reteach

Inequalities

A mathematical sentence that compares quantities is an **inequality**. Inequalities contain the symbols $<, >, \leq, \geq$.

$<$	$>$	$\leq$	$\geq$
• is less than • is fewer than	• is greater than • is more than	• is less than or equal to • is at most	• is greater than or equal to • is at least

Example 1 **Of the numbers 5, 6, or 7, which is a solution of the inequality $f + 4 < 10$?**

Value of f	$f + 4 < 10$	True or False
5	$5 + 4 < 10$ $9 < 10$	true
6	$6 + 4 < 10$ $10 < 10$	false
7	$7 + 4 < 10$ $11 < 10$	false

The number 5 makes a true sentence.

Example 2 **Is the given value a solution of the inequality?**

a. $x + 4 > 8, x = 5$

$x + 4 > 8$ Write the inequality.

$5 + 4 \overset{?}{>} 8$ Replace x with 5.

$9 \overset{?}{>} 8$ Simplify.

Since $9 > 8$, 5 is a solution.

b. $10 \leq 15 - y, y = 7$

$10 \overset{?}{\leq} 15 - 7$ Write the inequality, replacing y with 7.

$10 \overset{?}{\leq} 8$ Simplify.

Since 10 is not less than or equal to 8, 7 is not a solution.

Exercises

Determine which number is a solution of the inequality.

1. $7 + a > 13$; 5, 6, 7

2. $12 - b \leq 4$; 6, 7, 8

3. $9 + n \geq 20$; 9, 10, 11

Is the given value a solution of the inequality?

4. $y - 3 < 5, y = 9$

5. $14 + s \geq 22, s = 8$

6. $r - 5 > 6, y = 10$

NAME ______________________ DATE ____________ PERIOD _____

Skills Practice

Inequalities

Determine which number is a solution of the inequality.

1. $18 + a > 21$; 2, 3, 4

2. $24 - x \leq 19$; 3, 4, 5

3. $7 + r \geq 18$; 11, 10, 9

4. $9 - h > 2$; 6, 7, 8

5. $32 - n \leq 17$; 13, 14, 15

6. $16 + j \geq 29$; 13, 12, 11

7. $10 - f < 7$; 2, 3, 4

8. $52 + q < 56$; 5, 4, 3

Is the given value a solution of the inequality?

9. $2 + s \geq 10$; $s = 7$

10. $14 - r < 9$; $r = 6$

11. $j - 11 \geq 20$; $j = 32$

12. $t + 6 > 40$; $t = 35$

13. $16 + m > 40$; $m = 16$

14. $9x \geq 80$; $x = 9$

15. $15 \leq 3z$; $z = 4$

16. $2n > 26$; $n = 5$

NAME ______________________ DATE ____________ PERIOD _____

Reteach

Problem-Solving Investigation: Guess, Check, and Revise

When solving problems, one strategy that is helpful to use is *guess, check, and revise*. Based on the information in the problem, you can make a guess of the solution. Then use computations to check if your guess is correct. You can repeat this process until you find the correct solution.

You can use *guess, check, and revise*, along with the following four-step problem solving plan to solve a problem.

1 Understand – Read and get a general understanding of the problem.

2 Plan – Make a plan to solve the problem and estimate the solution.

3 Solve – Use your plan to solve the problem.

4 Check – Check the reasonableness of your solution.

Example

SPORTS **Patrice made a combination of 2-point baskets and 3-point baskets in the basketball game. She scored a total of 9 points. How many 2-point baskets and 3-point baskets did Patrice make in the basketball game?**

Understand You know that she made both 2-point and 3-point baskets. You also know she scored a total of 9 points. You need to find how many of each basket she made.

Plan Make a guess until you find an answer that makes sense for the problem.

Solve

Number of 2-Point Baskers	Number of 3-Point Baskers	Total Number of Points	Compare to 9
1	2	$1(2) + 2(3) = 8$	< 9
2	2	$2(2) + 2(3) = 10$	> 9
2	1	$2(2) + 1(3) = 7$	< 9
3	1	$3(2) + 1(3) = 9$	$= 9$

Check Three 2-point baskets result in 6 points. One 3-point basket results in 3 points. Since $6 + 3$ is 9, the answer is correct.

Exercise

VIDEO GAMES Blaine has 16 video games. The types of video games he has are sports games, treasure hunts, and puzzles. He has 4 more sports games than treasure hunts. He has 3 fewer puzzles than treasure hunts. Use the *guess, check, and revise* strategy to determine how many of each type of video game Blaine has.

NAME ______________________ DATE ____________ PERIOD _____

Skills Practice

Problem-Solving Investigation: Guess, Check, and Revise

Use the *guess, check, and revise* strategy to solve each problem.

1. **MONEY** Keegan has 10 coins in his pocket that total \$2.05. He only has quarters and dimes. How many of each coin does Keegan have?

2. **NUMBERS** Ms. Junkin told her students that she was thinking of three different numbers between 1 and 9 that had a sum of 19. Find three possible numbers.

3. **SHOPPING** Natasha bought some bracelets and some rings during a jewelry store sale. Each bracelet cost \$4 and each ring cost \$7. If Natasha spent \$29 on the jewelry, how many bracelets and rings did she buy?

4. **ORDER OF OPERATIONS** Use each of the symbols +, −, and × to make the following math sentence true.

   5 ___ 2 ___ 6 ___ 9 = 13

NAME ____________________ DATE __________ PERIOD ____

Reteach

Write and Graph Inequalities

Example 1 **Write an inequality for the sentence.**

Fewer than 70 students attended the last dance.

Words	*Fewer than* 70 students attended the last dance.
Variable	Let s = the number of students.
Inequality	$s < 70$

Example 2 **Graph each inequality on a number line.**

a. $x > 8$

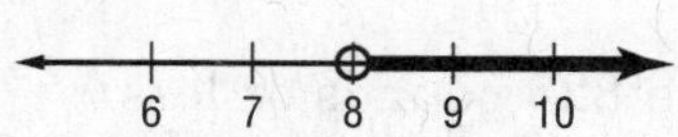

The **open dot** means 8 does *not* make the sentence true. The graph to the right means that numbers greater than 8 make the sentence true.

b. $x \leq 8$

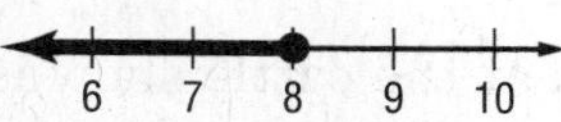

The **closed dot** means 8 *does* make the sentence true. The graph to the left means that numbers less than 8 make the sentence true.

Exercises

Write an inequality for each sentence.

1. The maximum height h is 45 feet.

2. The weight m of all adult male elephants is over 12,000 pounds.

3. The maximum fee f for any student is $15.

4. You must be at least 38 inches tall to ride the roller coaster.

Graph each inequality on the number line.

5. $x > 7$

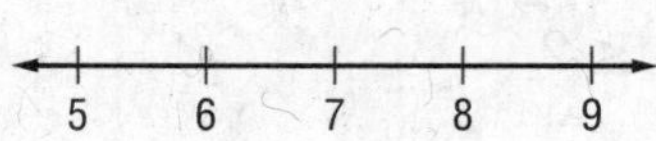

6. $s \geq 5$

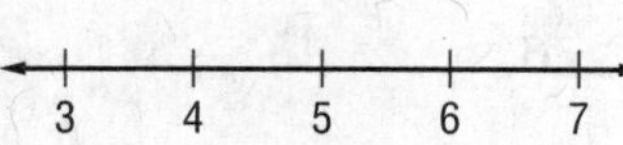

7. $m < 2$

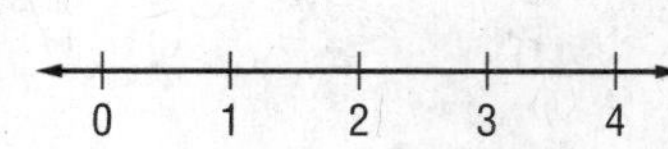

8. $f \leq 6$

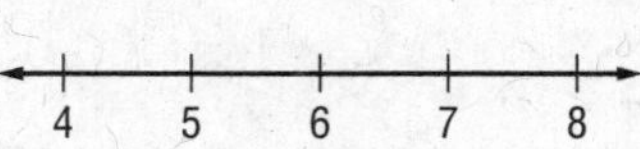

9. $r > 11$

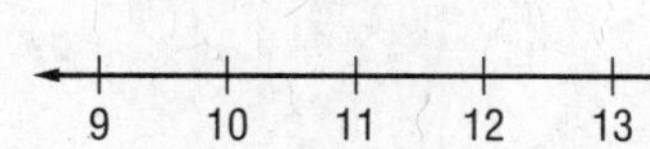

10. $n \geq 3$

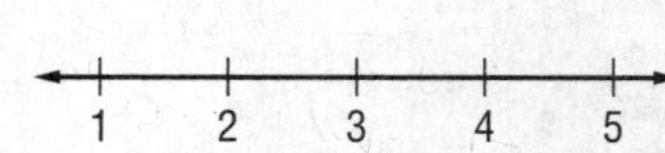

NAME ______________________ DATE ____________ PERIOD _____

Skills Practice

Write and Graph Inequalities

Write an inequality for each sentence.

1. More than 40,000 fans attended the opening football game at the University of Florida.

2. Her earnings were no more than $86.

3. A savings account balance is now less than $550.

4. The number of club members is at least 25.

5. The spring calf class at the cattle show is for calves that weigh 825 pounds or less.

6. The minimum deposit for a new checking account is $75.

Graph each inequality on the number line.

7. $a < 8$

6 7 8 9 10

8. $d \leq 4$

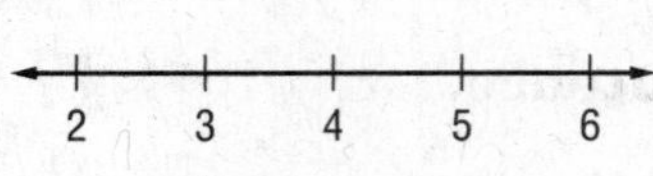

9. $b \geq 11$

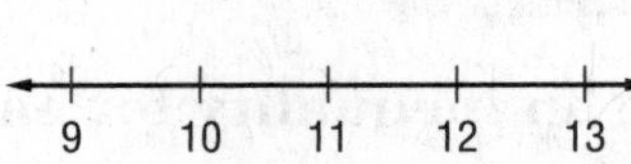

10. $x > 3$

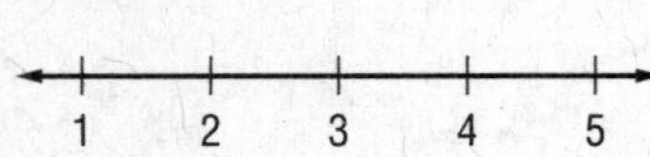

11. $x < 9$

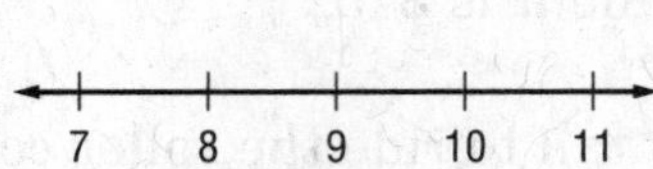

12. $r \geq 4$

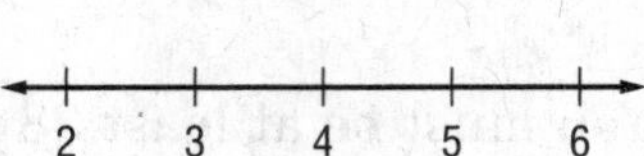

13. $x > 10$

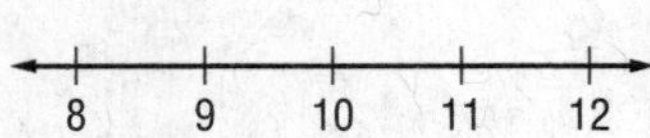

14. $x \geq 6$

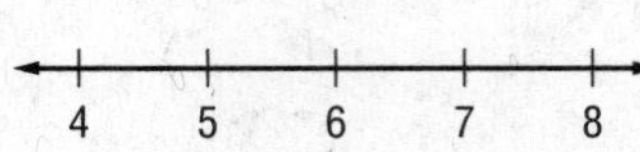

15. $x \leq 2$

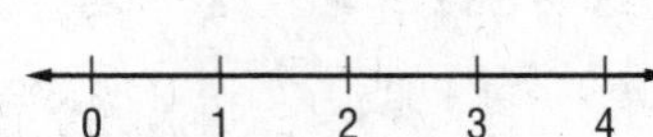

Multi-Part Lesson 2

PART F

NAME ______________________ DATE ____________ PERIOD _____

Reteach

Solve One-Step Inequalities

Addition and Subtraction Properties	
Words	When you add or subtract the same number from each side of an inequality, the inequality remains true.
Symbols	For all numbers a, b, and c, 1. if $a < b$, then $a + c < b + c$ and $a - c < b - c$. 2. if $a > b$, then $a + c > b + c$ and $a - c > b - c$.

Example 1 **Solve $x + 9 \leq 12$. Graph the solution on a number line.**

$x + 9 \leq 12$ Write the inequality.

$-9 \quad -9$ Subtract 9 from each side.

$x \leq 3$ Simplify.

The solution is $x \leq 3$. To graph it, draw a closed dot at 3 and draw an arrow to the left on the number line.

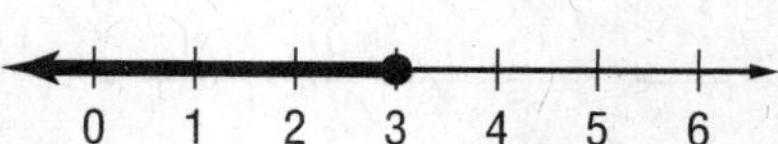

Multiplication and Division Properties	
Words	When you multiply or divide each side of an inequality by the same *positive* number, the inequality remains true.
Symbols	For all numbers a, b, and c, where $c > 0$, 1. if $a < b$, then $ac < bc$ and $\frac{a}{c} < \frac{b}{c}$. 2. if $a > b$, then $ac > bc$ and $\frac{a}{c} > \frac{b}{c}$.

Example 2 **Solve $3x > 15$. Graph the solution on a number line.**

$3x > 15$ Write the inequality.

$\frac{3x}{3} > \frac{15}{3}$ Divide each side by 3.

$x > 5$ Simplify.

The solution is $x > 5$. To graph it, draw an open dot at 5 and draw an arrow to the right on the number line.

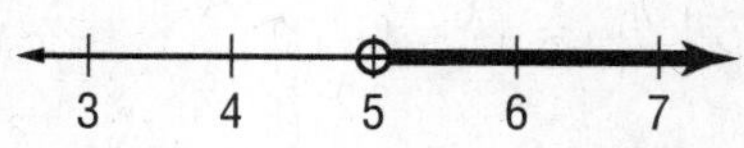

Exercises

Solve each inequality. Then graph the solution on a number line.

1. $9d > 81$

2. $t - 5 < 4$

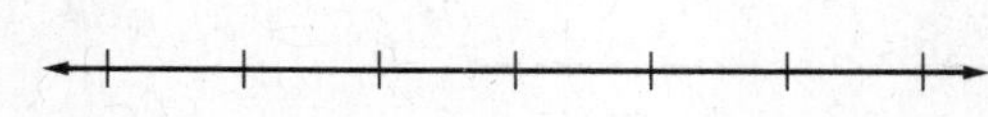

3. $j + 6 \geq 11$

4. $\frac{n}{3} \leq 7$

NAME ______________________________ DATE ____________ PERIOD _____

Skills Practice

Solve One-Step Inequalities

Solve each inequality. Graph the solution on a number line.

1. $8x > 16$

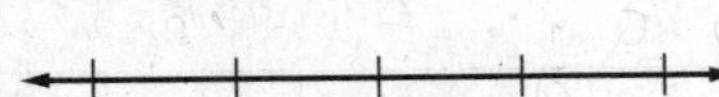

2. $h - 9 > 13$

3. $t + 7 \leq 12$

4. $\frac{r}{3} \geq 5$

5. $j + 4 < 10$

6. $7y < 35$

7. $b - 15 > 11$

8. $\frac{n}{4} < 4$

9. $4b \geq 12$

10. $z + 8 \geq 14$

11. $\frac{y}{6} \geq 2$

12. $j - 8 < 9$

13. $k - 10 \geq 6$

14. $6z \geq 18$

15. $12a \geq 48$

16. $\frac{s}{3} < 5$

17. SHOPPING Chantal would like to buy a new pair of running shoes. The shoes she likes start at \$85. If she has already saved \$62, write an inequality to show how much more money Chantal must save.

NAME ______________________ DATE ____________ PERIOD _____

Reteach

Two-Step Inequalities

Example **Four more than 3 times a number is less than 28. Write, solve, and graph an inequality to represent this situation.**

Words	*Four more than 3 times a number is less than 28.*
Variable	Let n = the number.
Inequality	$3n + 4 < 28$

$3n + 4 < 28$ Write the inequality.

$3n + 4 < 28$
$\underline{\quad -4 \quad -4}$ Subtract 4 from each side.

$3n < 24$ Simplify.

$\frac{3n}{3} < \frac{24}{3}$ Divide each side by 3.

$n < 8$ Simplify.

The solution is $n < 8$. To graph it, draw an open circle at 8 and draw an arrow to the left on the number line.

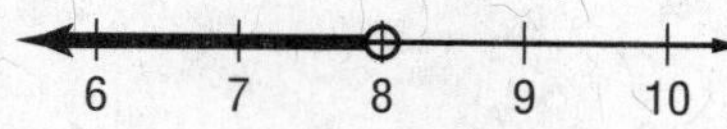

Exercises

Write, solve, and graph an inequality to represent each sentence.

1. 25 is less than 10 more than 5 times a number.

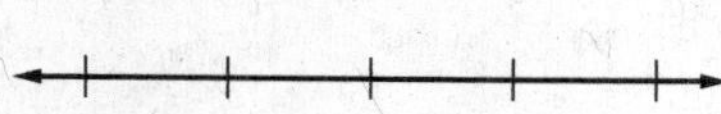

2. Eight less than 3 times a number is greater than or equal to 10.

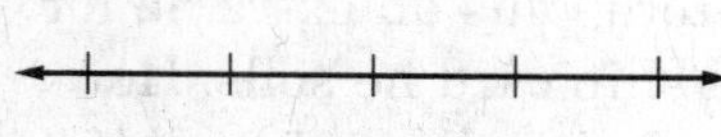

3. Five more than 6 times a number is less than or equal to 29.

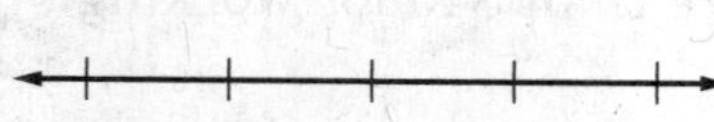

4. **PEDAL BOATS** It costs \$25 plus \$6 per hour to rent a pedal boat at the lake. The minimum fee is \$43. What is the minimum number of hours for which you can rent the pedal boat?

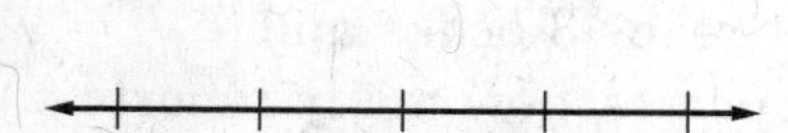

NAME ______________________ DATE ____________ PERIOD ______

Skills Practice

Two-Step Inequalities

Solve and graph each inequality.

1. $12 > 6 + 2n$

2. $4n - 7 \geq 21$

3. $5n - 16 > 29$

4. $24 \leq 12 + 3n$

5. $2n - 9 \geq 17$

6. $56 > 6 + 10n$

7. **SALES** Eric gets paid $7 for each hour he works at the electronics store, plus an extra $2 for every store membership card he sells. How many memberships does he have to sell if he wants to make more than $40 for working 4 hours?

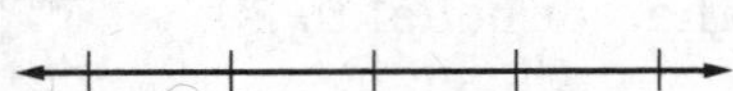

8. **CONTEST** In a class contest, students get 8 points for each book they read plus an extra 5 points for each book report they write. If a student reads 3 books and wants at least 34 points, how many reports does she have to write?

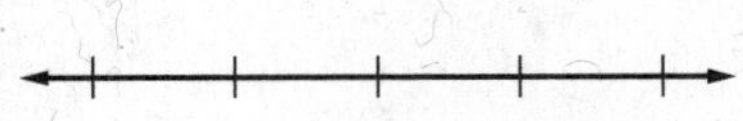

NAME ______________________ DATE ____________ PERIOD ______

Reteach

Integers and Absolute Value

An **integer** is a number from the set {…, –4, –3, –2, –1, 0, 1, 2, 3, 4, …}. Integers greater than 0 are **positive integers**. Integers less than 0 are **negative integers**. Always use the negative sign (–) to indicate a negative number.

Example 1 **Write an integer for each situation.**

a. 16 feet under the ground
Because it is *under* the ground, the integer is –16.

b. a gain of 5 hours
Because it is a *gain*, the integer is 5.

Example 2 **Graph the set of integers {–5, –2, 3} on a number line.**

Draw a number line. Draw a dot at the location of –5, of –2, and of 3.

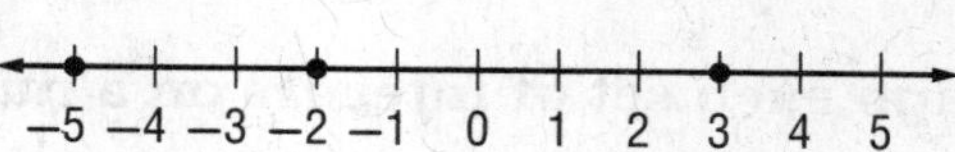

The **absolute value** of a number is the distance between the number and zero on a number line. The symbol for absolute value is | |.

Example 3 **Evaluate each expression.**

a. $|-2|$

The graph of -2 is 2 units from 0.

So, $|-2| = 2$.

2 units

–3 –2 –1 0 1 2 3

b. $|8| + |-6|$

$|8| + |-6| = 8 + |-6|$ The absolute value of 8 is 8.

$= 8 + 6$ The absolute value of -6 is 6.

$= 14$ Simplify.

Exercises

Write an integer for each situation.

1. a profit of \$60

2. a decrease of 10°

3. Graph the set {–6, 5, –4} on a number line.

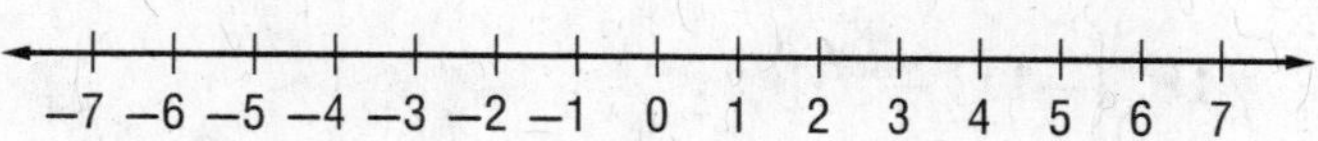

4. Evaluate $|-11| - |5|$.

5. Evaluate $|-1| + |-7|$.

NAME ______________________ DATE ____________ PERIOD ______

Multi-Part Lesson 3 PART B

Skills Practice

Integers and Absolute Value

Write an integer for each situation.

1. a $5 discount

2. a growth of 2.5 centimeters

3. rise of 1,000 feet

4. loss of 6 pounds

5. price drop of $12

6. pay raise of $1 per hour

Graph each set of integers on a number line.

7. {−2, 0, 4}

−5 −4 −3 −2 −1 0 1 2 3 4 5

8. {7, −3, −1}

−3 −2 −1 0 1 2 3 4 5 6 7

9. {5, −5, −6, 2}

−7 −6 −5 −4 −3 −2 −1 0 1 2 3 4 5 6 7

10. {−8, −7, −4, 1}

−9 −8 −7 −6 −5 −4 −3 −2 −1 0 1 2 3

11. {−11, −7, 3, 6}

−12 −10 −8 −6 −4 −2 0 2 4 6

12. {−10, −8, −14, 8}

−12 −9 −6 −3 0 3 6 9

Evaluate each expression.

13. $|-4|$

14. $|800|$

15. $|10 - 5|$

16. $|-18| - |13|$

17. $|16| - |-2|$

18. $|-11| + |0|$

19. $|-26| + |-17|$

20. $|-1| + |-1|$

Multi-Part Lesson **3** PART **C**

NAME _______________ DATE _______ PERIOD _____

Reteach

The Coordinate Plane

The *x*-axis and *y*-axis separate the coordinate plane into four regions called **quadrants**.

Example 1 **Identify the ordered pair that names Point *A*.**

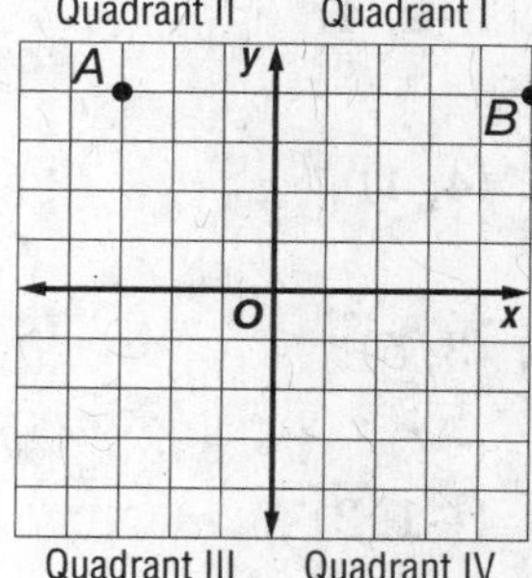

Step 1 Start at the origin. Move left on the *x*-axis to find the *x*-coordinate of point *A*, which is −3.

Step 2 Move up the *y*-axis to find the *y*-coordinate, which is 4. Point *A* is named by (−3, 4).

Example 2 **Graph Point *B* at (5, 4).**

Step 1 Use the coordinate plane shown above. Start at the origin. The *x*-coordinate is 5, so move 5 units to the right.

Step 2 Since the *y*-coordinate is 4, move 4 units up.

Step 3 Draw a dot. Label the dot *B*.

Exercises

Use the coordinate plane at the right. Write the ordered pair that names each point.

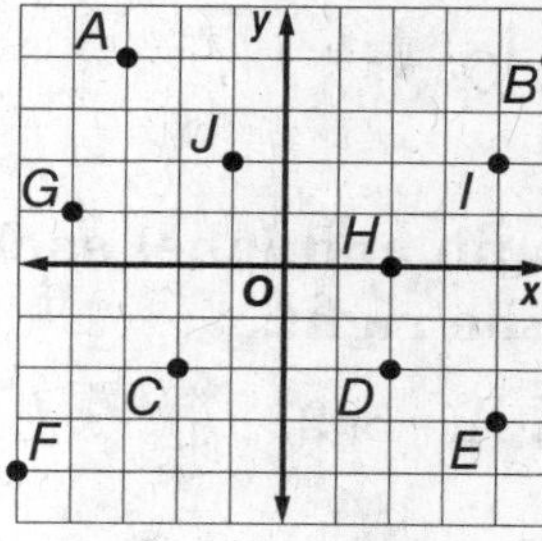

1. *C*

2. *D*

3. *E*

4. *F*

5. *G*

6. *H*

7. *I*

8. *J*

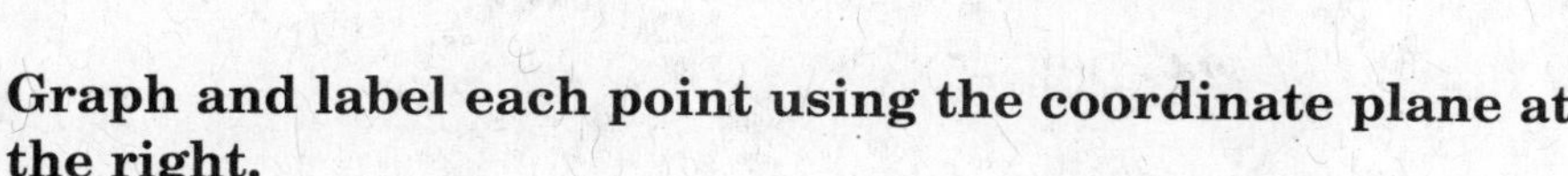

Graph and label each point using the coordinate plane at the right.

9. *R*(−2, 3)

10. *P*(3, −2)

11. *Z*(−1, 0)

12. *B*(−3, −4)

13. *S*(4, 1)

14. *M*(1, −3)

NAME ______________________ DATE __________ PERIOD _____

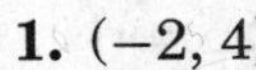

Multi-Part Lesson 3 PART C

Skills Practice

The Coordinate Plane

Use the coordinate plane at the right. Identify the point for each ordered pair.

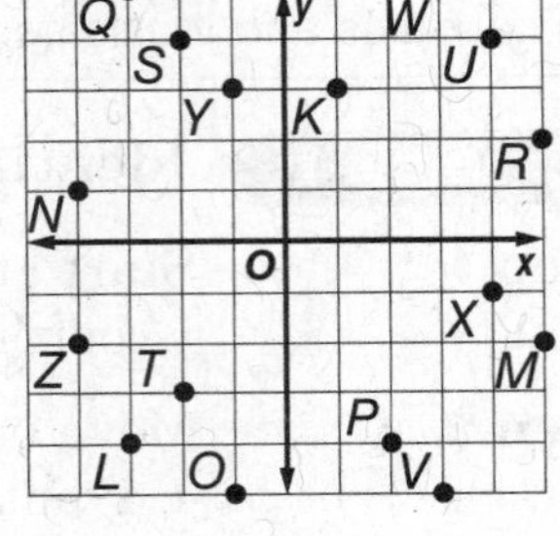

1. (−2, 4)
2. (−2, −3)
3. (4, 4)
4. (3, −5)
5. (3, 5)
6. (4, −1)
7. (−1, 3)
8. (−4, −2)

Use the coordinate plane above. Write the ordered pair that names each point. Then identify the quadrant where each point is located.

9. *K*
10. *L*
11. *M*
12. *N*
13. *O*
14. *P*
15. *Q*
16. *R*

Graph and label each point on the coordinate plane at the right.

17. *A*(−5, 2)
18. *I*(2, 1)
19. *J*(1, −3)
20. *B*(−5, −1)
21. *C*(3, 3)
22. *K*(−1, 2)
23. *L*(0, −1)
24. *D*(2, −5)
25. *E*(3, −2)
26. *M*(−4, −5)
27. *N*(1, 5)
28. *F*(−2, 5)
29. *G*(−1, −4)
30. *O*(5, −5)

NAME ______________________ DATE ____________ PERIOD ______

Reteach

Points, Lines, and Planes

Example 1 **Identify the figure below. Then name it using symbols.**

The figure has two endpoints without arrows. So, it is a line segment.

symbols: $\overline{CD}$ or $\overline{DC}$

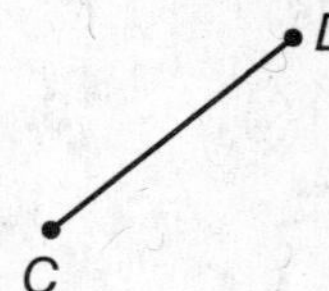

Example 2 **Measure each segment. Then determine whether the line segments are congruent.**

$\overline{RS}$ is 3.5 centimeters and $\overline{PQ}$ is 2 centimeters, so they are *not* congruent.

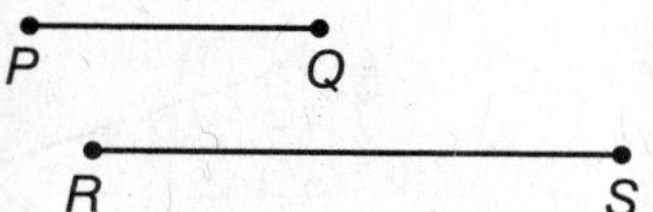

Example 3 FLAGS **Classify the lines formed on the flag at the right as *intersecting*, *perpendicular*, or *parallel*. Choose the most specific term.**

The lines are the same distance apart and never intersect. So, they are parallel lines.

Exercises

Identify each figure. Then name it using symbols.

1. 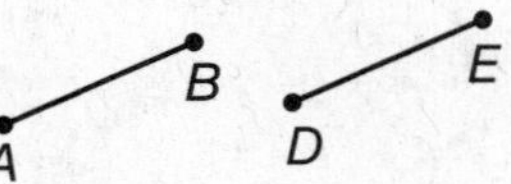

2. • G

3. T U

Measure each segment. Then determine whether each pair of line segments are congruent. Write *yes* or *no*.

4. A B D E

5. 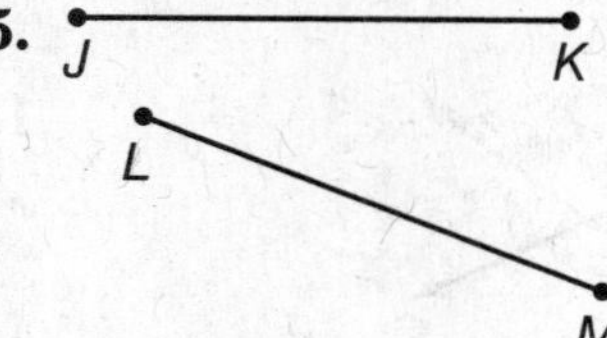

6. 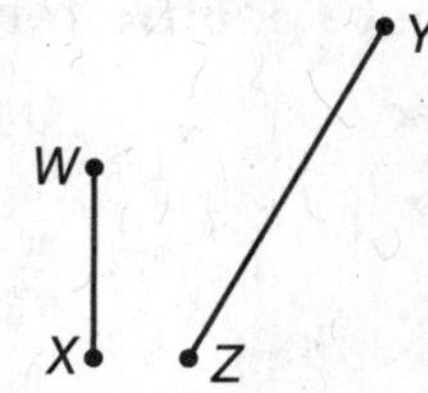

Classify each pair of lines as *intersecting*, *perpendicular*, or *parallel*. Choose the most specific term.

7. 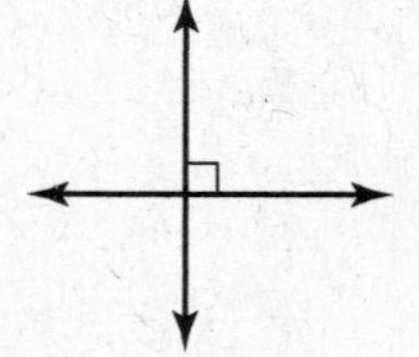

8. 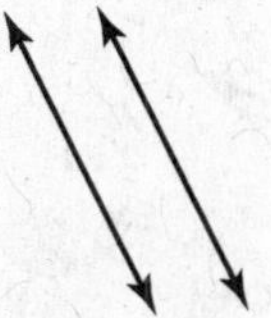

9. 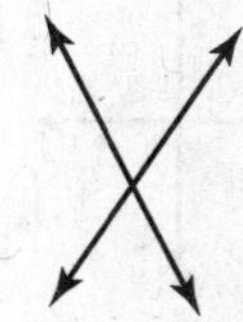

NAME ______________________________ DATE ____________ PERIOD ______

Skills Practice

Points, Lines, and Planes

Exercises

Identify each figure. Then name it using symbols.

1. V W

2. B C

3. M N

4. K L

5. E F

6. *D*

Measure each segment. Then determine whether each pair of line segments are congruent. Write *yes* or *no*.

7.

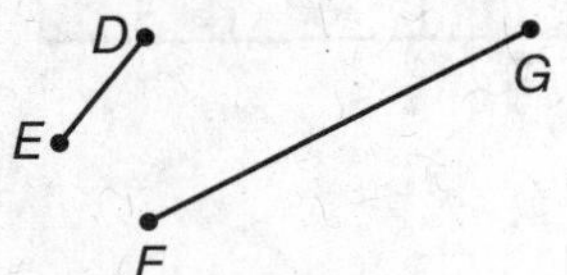

8.

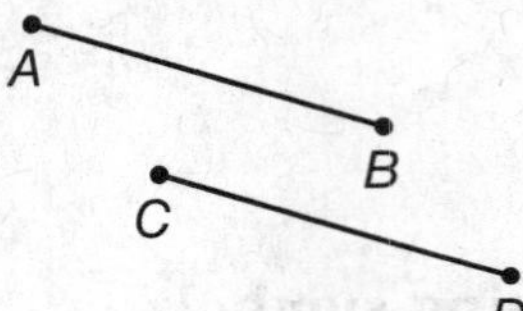

9.

10.

11.

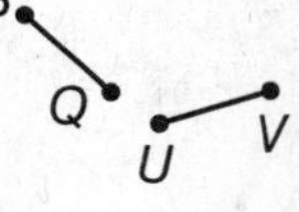

12.

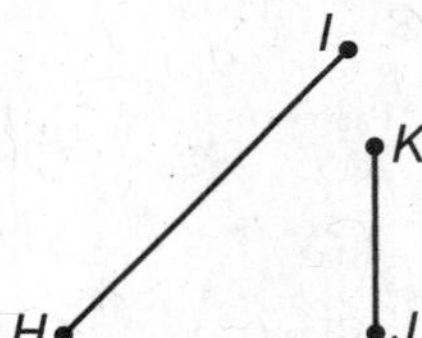

Classify each pair of lines as *intersecting*, *perpendicular*, or *parallel*. Choose the most specific term.

13.

14.

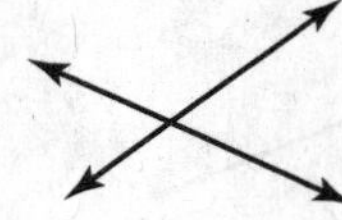

15.

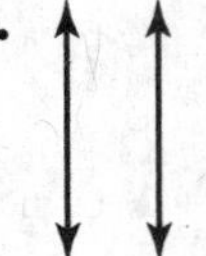

16.

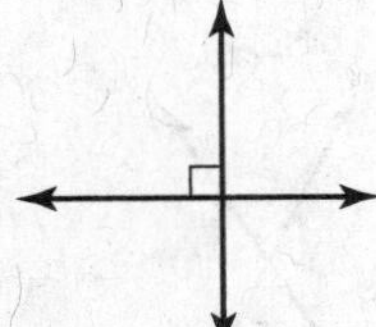

17.

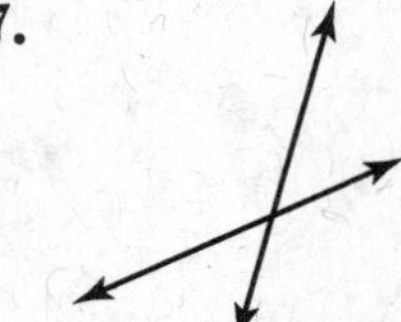

18.

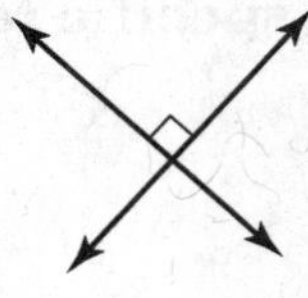

NAME _______________ DATE __________ PERIOD ____

Reteach

Measure Angles

Example 1 **Use a protractor to find the measure of the angle.**

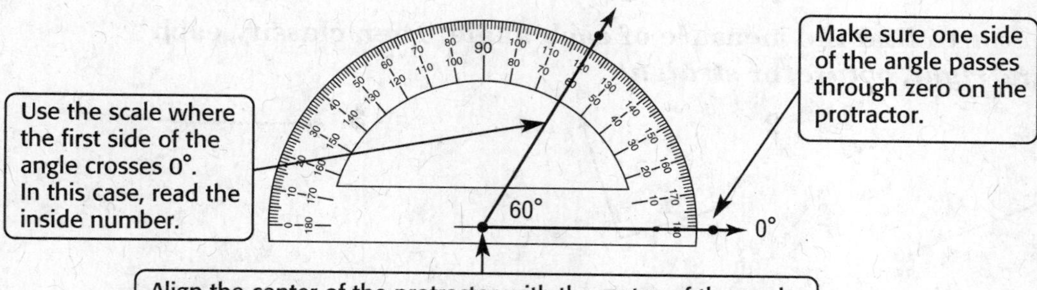

The angle measures 60°.

Example 2 **Classify the angle as *acute*, *right*, *obtuse*, or *straight*.**

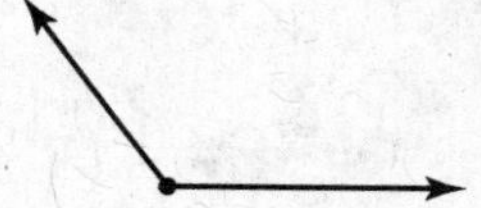

The angle is between 90° and 180°, so it is an obtuse angle.

Example 3 **Classify the angle as *acute*, *right*, *obtuse*, or *straight*.**

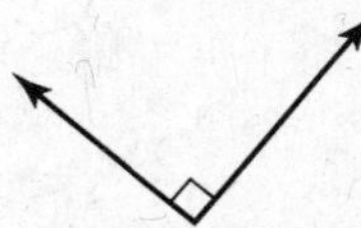

The angle is exactly 90°, so it is a right angle.

Exercises

Use a protractor to find the measure of each angle. Then classify the angle as *acute*, *right*, *obtuse*, or *straight*.

1.

2.

3.

4.

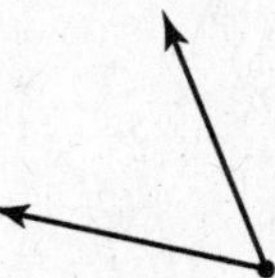

5.

6.

Multi-Part Lesson 1 PART B

NAME ______________________________ DATE ____________ PERIOD ______

Skills Practice

Measure Angles

Exercises

Use a protractor to find the measure of each angle. Then classify each angle as *acute*, *right*, *obtuse*, or *straight*.

1. 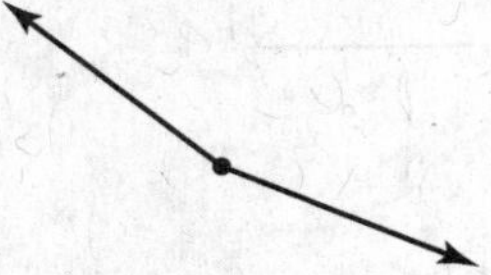

2. 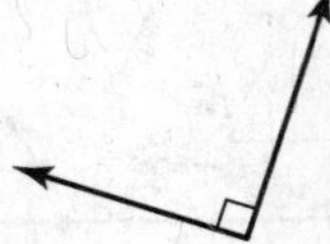

3. 

4. 

5. 

6. 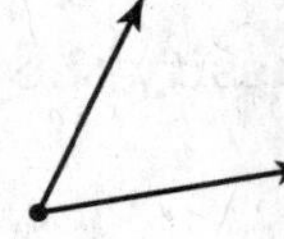

7. 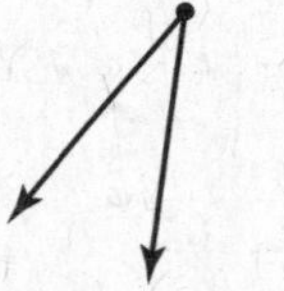

8. 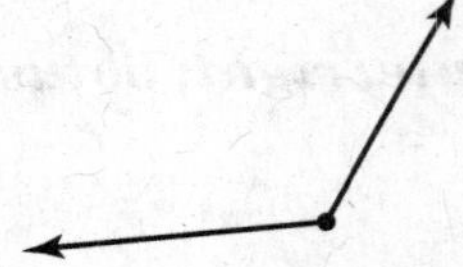

9. 

10. 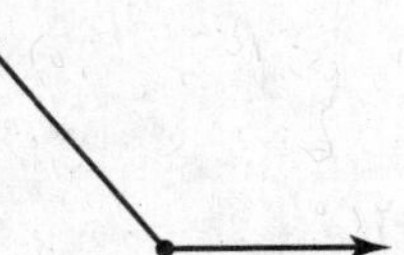

11. 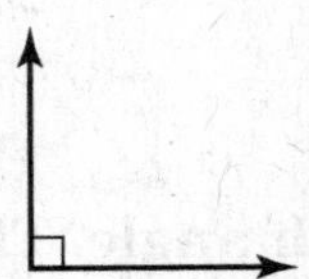

12. 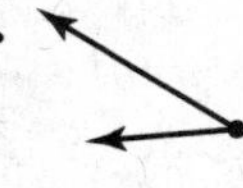

13. 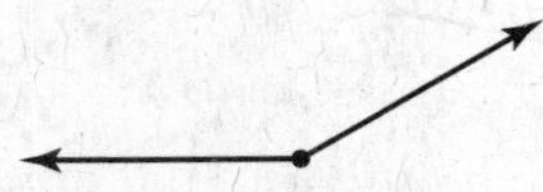

14. 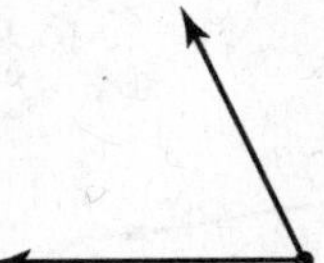

15. 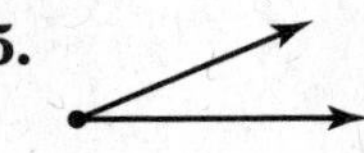

NAME ______________________________ DATE ____________ PERIOD ____

Reteach

Angle Relationships

Vertical angles are the opposite angles formed by intersecting lines. Vertical angles are **congruent angles**, or angles with the same measures.

Example 1 **Find the value of x in the figure at the right.**

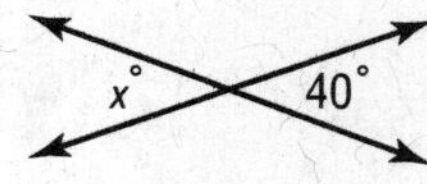

The angle labeled $x°$ and the angle labeled 40° are vertical angles. Therefore, they are congruent.

So, the value of x is 40.

Two angles are **complementary** if the sum of their measures is 90°.
Two angles are **supplementary** if the sum of their measures is 180°

Example 2 **Classify the pair of angles at the right as *complementary*, *supplementary*, or *neither*.**

130°
50°

$130° + 50° = 180°$
The angles are supplementary.

Example 3 **Find the value of x in the figure at the right.**

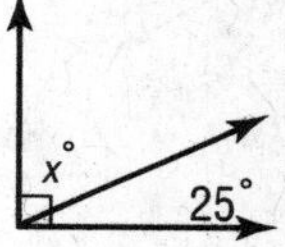

Since the angles form a right angle, they are complementary.

$x° + 25° = 90°$	Definition of complementary angles
$-25° = -25°$	Subtract 25 from each side.
$x° = 65°$	Simplify.

So, the value of x is 65.

Exercises

Classify each pair of angles as *complementary, supplementary,* or *neither*.

1. 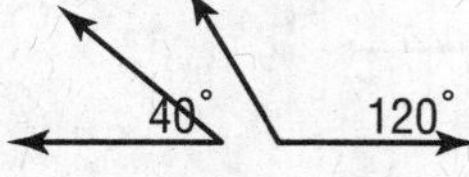

2. 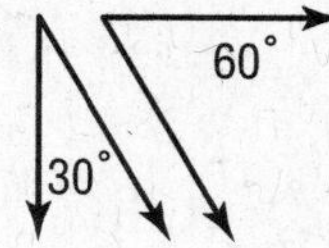

3. 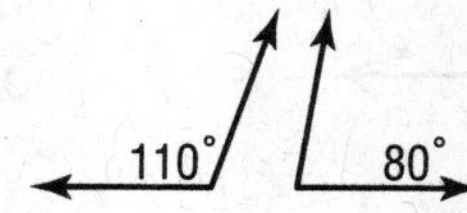

Find the value of x in each figure.

4. 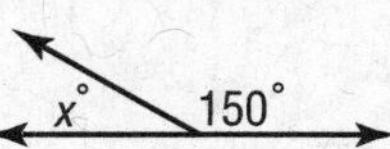

5. 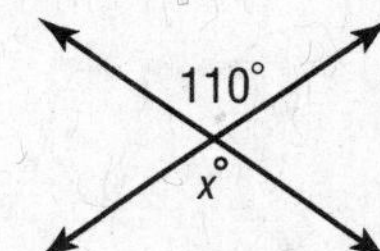

6. 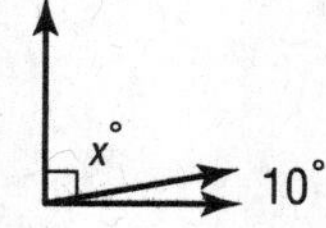

NAME ______________________ DATE ____________ PERIOD ______

Skills Practice

Angle Relationships

Classify each pair of angles as *complementary, supplementary,* or *neither*.

1.

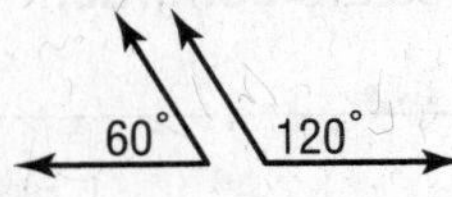

2.

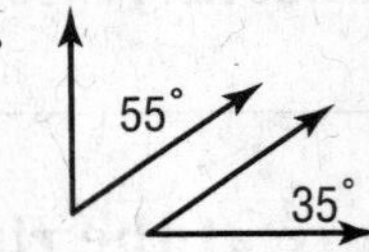

3.

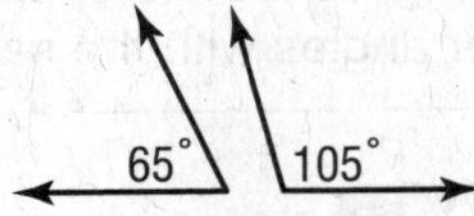

4.

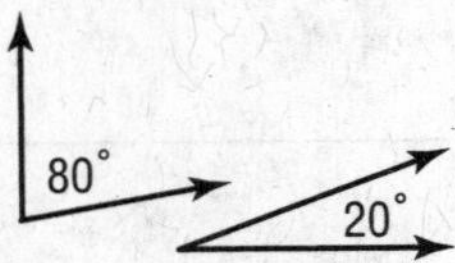

5.

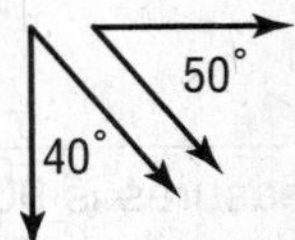

6.

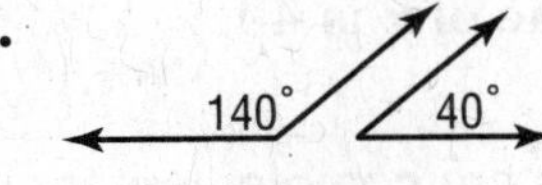

Find the value of x in each figure.

7.

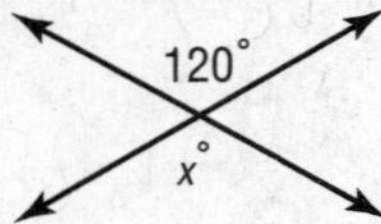

8.

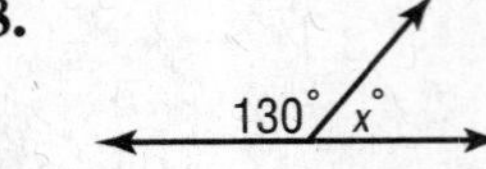

9.

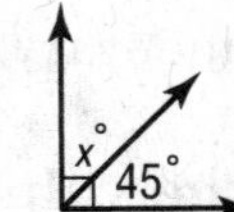

10.

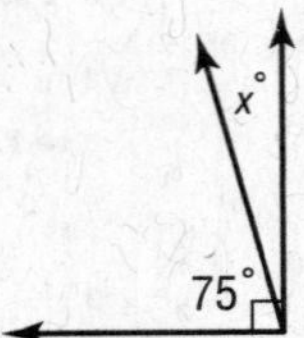

11.

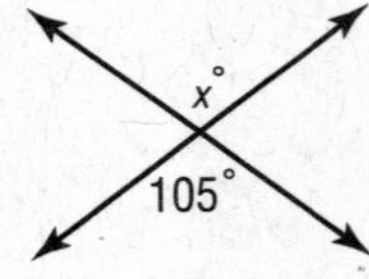

12.

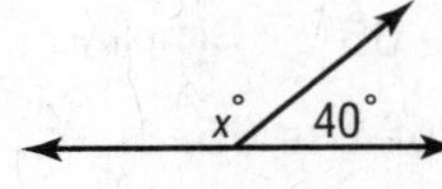

13.

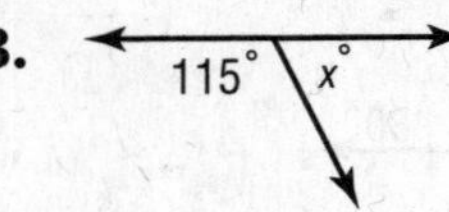

14.

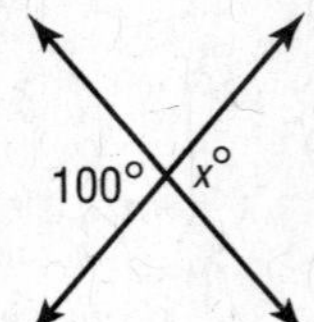

15.

x°
30°

Multi-Part Lesson 2 PART B

NAME ______________________ DATE ____________ PERIOD _____

Reteach

Triangles

Acute triangles have all acute angles. **Right triangles** have one right angle. **Obtuse triangles** have one obtuse angle. The sum of the angle measures in a triangle is 180°.

Example 1 **Classify the triangle at the right as *acute*, *right*, or *obtuse*.**

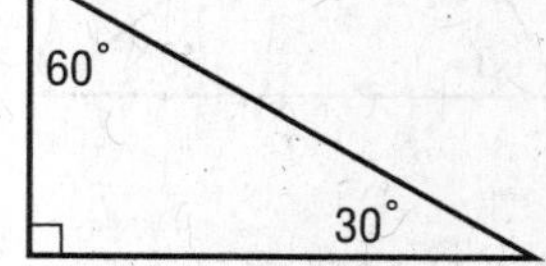

The triangle has one right angle.

So, the triangle is a right triangle.

Scalene triangles have no congruent sides. **Isosceles triangles** have at least 2 congruent sides. **Equilateral triangles** have 3 congruent sides.

Example 2 **Classify the triangle at the right as *scalene*, *isosceles*, or *equilateral*.**

Two of the sides measure 8 centimeters and are congruent.

So, the triangle is an isosceles triangle.

Example 3 **Find the value of x in the triangle at the right.**

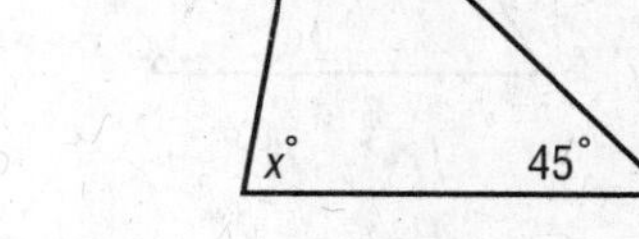

$x + 55 + 45 = 180$	Write the equation.
$x + 100 = 180$	Add 55 and 45.
$-100 = -100$	Subtract 100 from each side.
$x = 80$	

So, the value of x is 80.

Exercises

Classify each triangle as *acute*, *right*, or *obtuse*.

1.

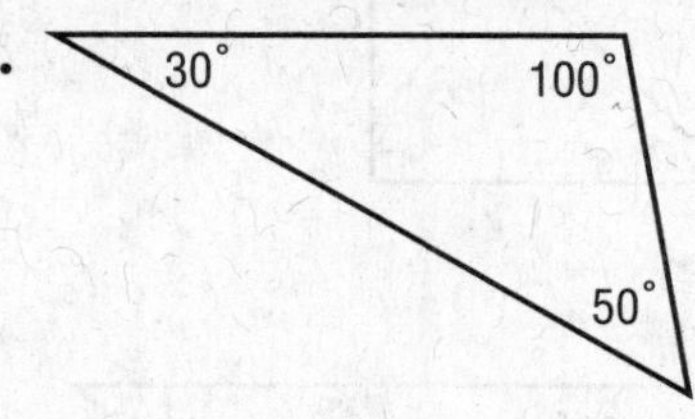

2.

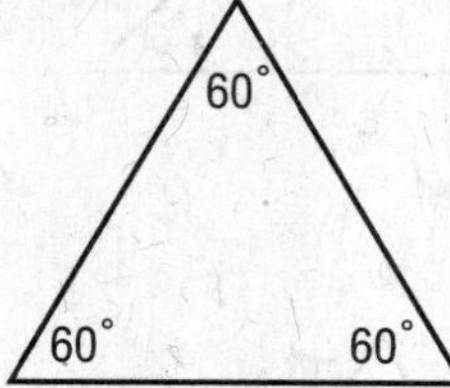

3.

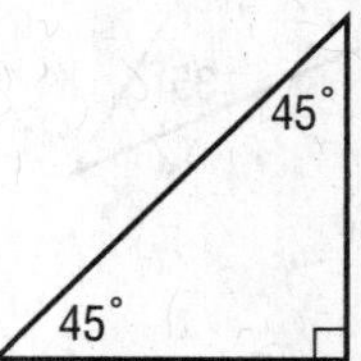

4. Find the value of x in the triangle at the right.

5. Classify the triangle at the right as *scalene*, *isosceles*, or *equilateral*.

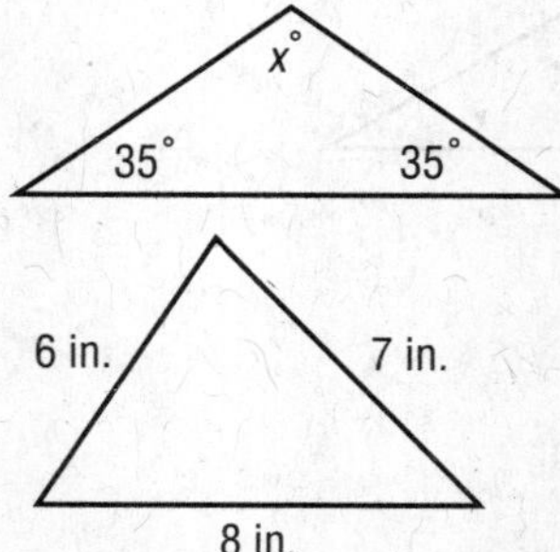

NAME ______________________ DATE ____________ PERIOD _____

Multi-Part Lesson 2 PART B

Skills Practice

Triangles

Classify each triangle as *acute, right,* or *obtuse.*

1.

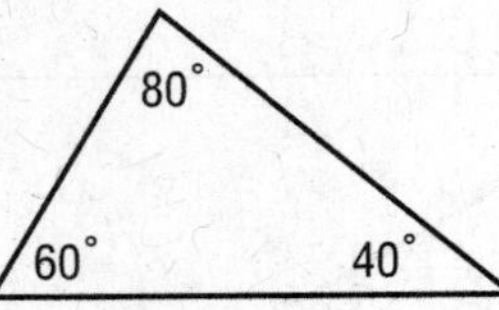

2.

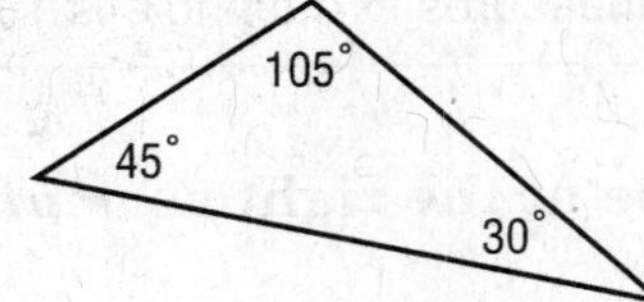

3.

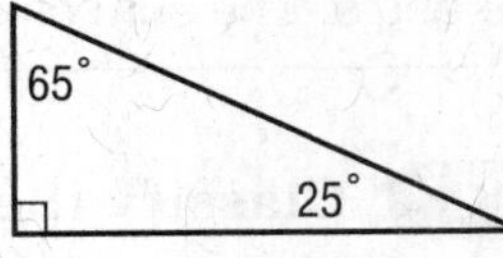

4.

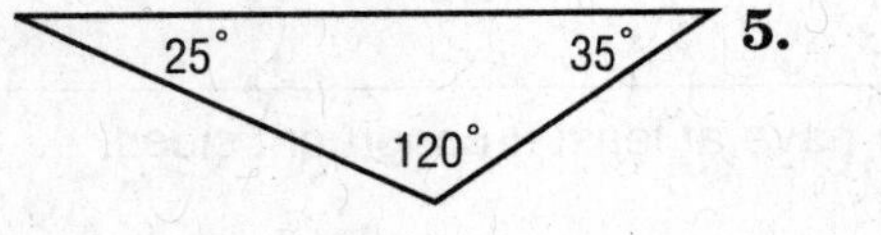

5.

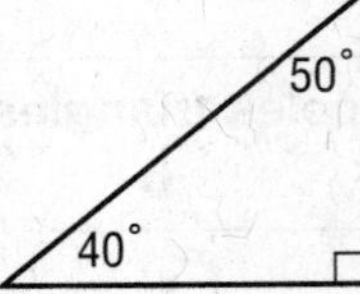

6.

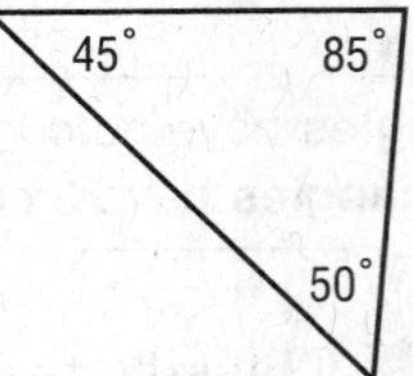

Classify each triangle as *scalene, isosceles,* or *equilateral.*

7.

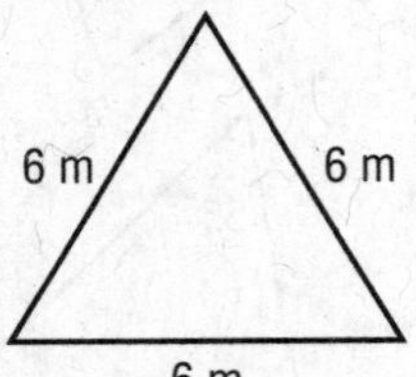

8.

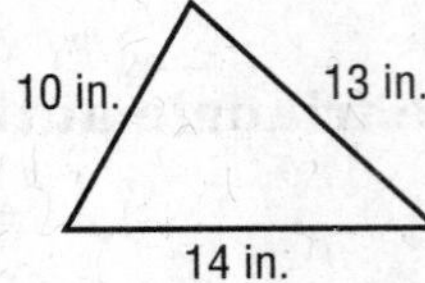

9.

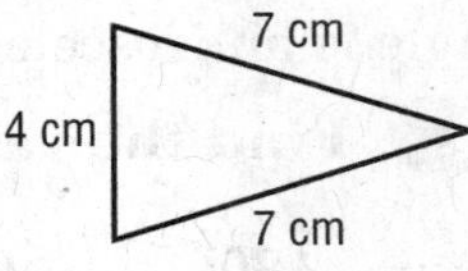

Find the value of *x* in each triangle.

10.

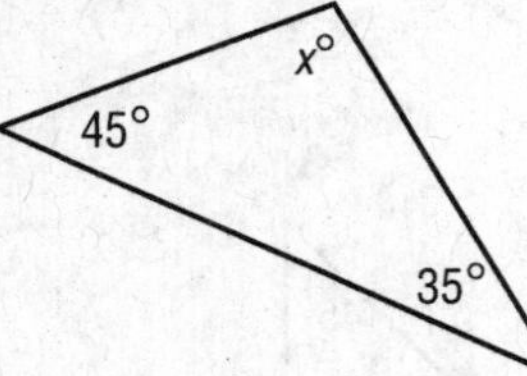

11.

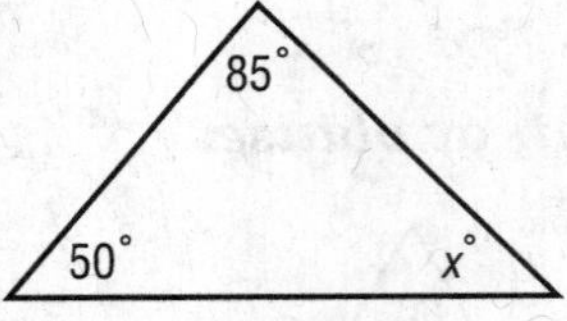

12.

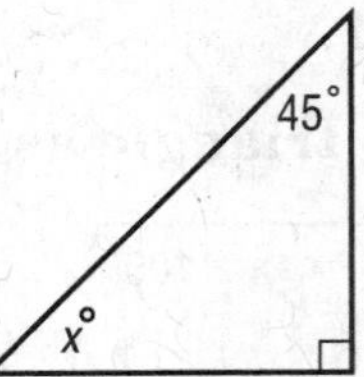

13.

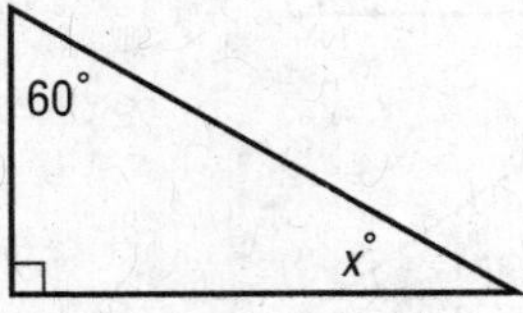

14.

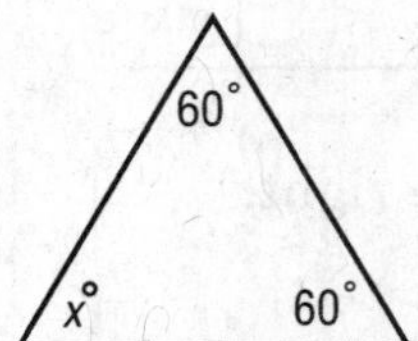

15.

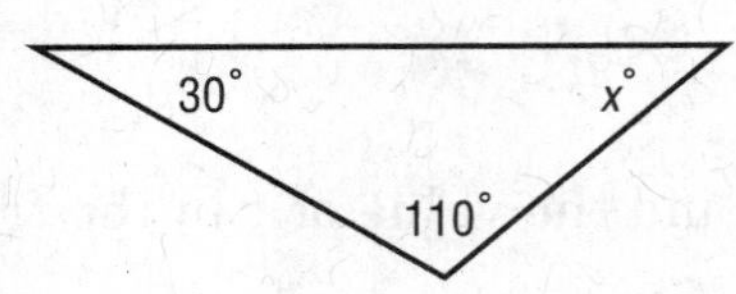

NAME ______________________ DATE ____________ PERIOD _____

Reteach

Problem-Solving Investigation: Draw a Diagram

When solving problems, one strategy that is helpful is to *draw a diagram*. A problem may often describe a situation that is easier to solve visually. You can draw a diagram of the situation, and then use the diagram to solve the problem.

You can draw a diagram, along with the following four-step problem solving plan to solve a problem.

1 Understand – Read and get a general understanding of the problem.

2 Plan – Make a plan to solve the problem and estimate the solution.

3 Solve – Use your plan to solve the problem.

4 Check – Check the reasonableness of your solution.

Example LIBRARY **The school library is putting tables in an open area that is 28 feet by 50 feet. Each table is a square with sides measuring 5 feet, and the tables must be 6 feet apart from each other and the wall. How many tables can fit in this area?**

Understand You know all the dimensions. You need to find how many tables will fit in this area.

Plan Draw a diagram to see how many tables will fit.

Solve

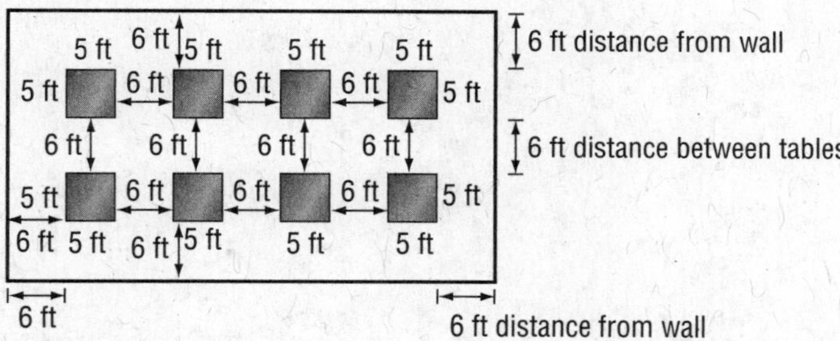

The diagram shows that 8 tables will fit in this area in the library.

Check Make sure the dimensions meet the requirements. The distance across is 50 feet and the distance down is 28 feet. So, the answer is correct.

Exercise

PICTURE FRAME La Tasha is decorating a picture frame by gluing gemstones around the frame. The picture frame is 7 inches by $5\frac{1}{2}$ inches. Each gemstone is $\frac{1}{2}$-inch wide and La Tasha glues them 1 inch apart and 1 inch from the edge. How many gemstones can La Tasha fit on the frame?

NAME ______________________ DATE ____________ PERIOD ____

Skills Practice

Problem-Solving Investigation: Draw a Diagram

Solve. Use the *draw a diagram* strategy.

1. **TRAVEL** Jasmine lives in Glacier and works in Alpine. There is no direct route from Glacier to Alpine, so Jasmine can drive through either Elm, Perth, or both towns to get to work. How many different ways can she drive to work?

2. **GARDENING** Ms. Kennedy is planting a vegetable garden in a rectangular area that is 3 feet by 6 feet. Each plant must be 12 inches from the other plants and the sides of the garden. How many vegetable plants can Ms. Kennedy plant in this rectangular garden?

3. **DRIVING** A downtown neighborhood is rectangular, 3 blocks by 5 blocks. How many ways are there to drive from one corner of the neighborhood to the opposite corner, if you must make exactly two turns?

NAME ________________________ DATE ____________ PERIOD _____

Reteach

Properties of Quadrilaterals

A **rectangle** has opposite sides congruent and parallel, and all right angles.
A **square** has all sides congruent, opposite sides parallel, and all right angles.
A **parallelogram** has opposite sides congruent and parallel, and opposite angles congruent.
A **rhombus** has all sides congruent, opposite sides parallel, and opposite angles congruent.
A **trapezoid** has exactly one pair of opposite sides parallel.

Example 1 **Classify the quadrilateral at the right.**

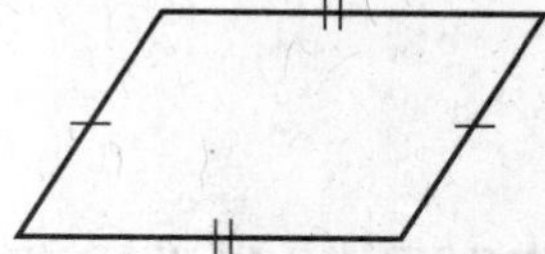

The figure has opposite sides congruent and parallel.

So, the figure is a parallelogram.

Quadrilaterals have four sides and four angles. The sum of the measures of the angles is 360°.

Example 2 **Find the value of x in the quadrilateral at the right.**

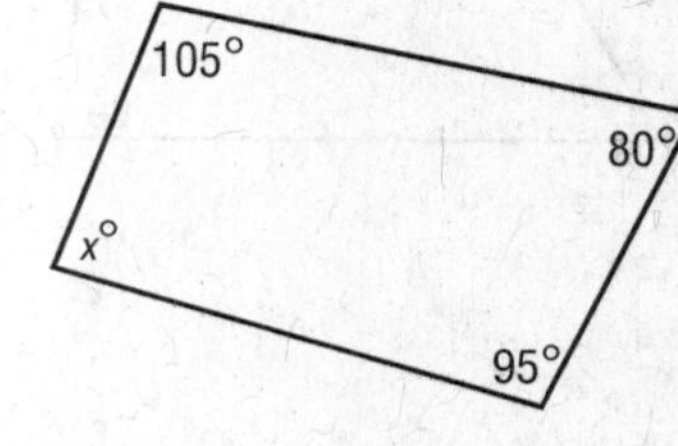

$x + 105 + 80 + 95 = 360$	Write the equation.
$x + 280 = 360$	Simplify: 105 + 80 + 95.
$-280 = -280$	Subtract 280 from each side.
$x = 80$	Simplify.

So, the value of x is 80.

Exercises

Classify each quadrilateral.

1. 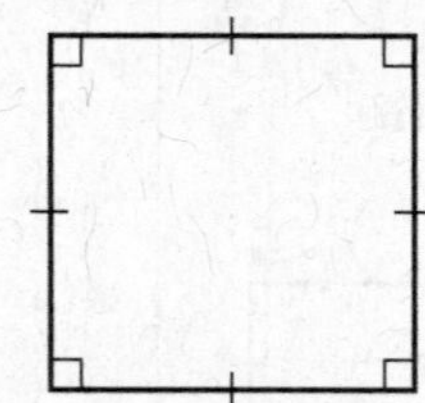

2. 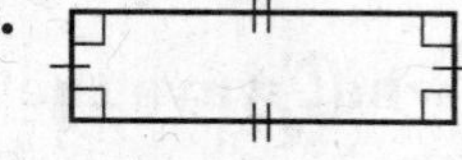

3. 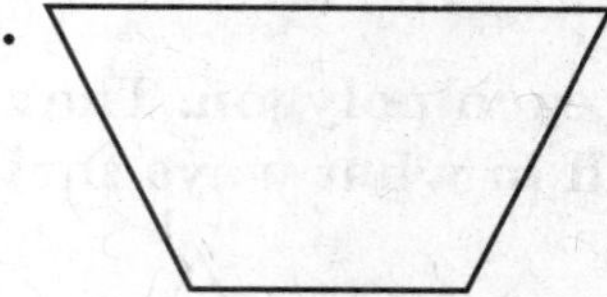

Find the value of x in each quadrilateral.

4. 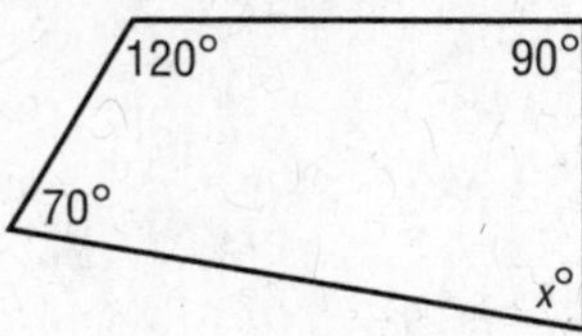

5. 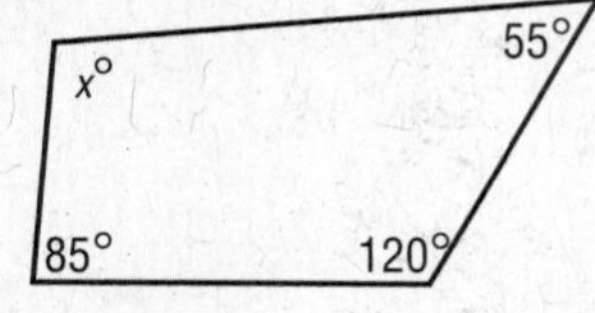

6. 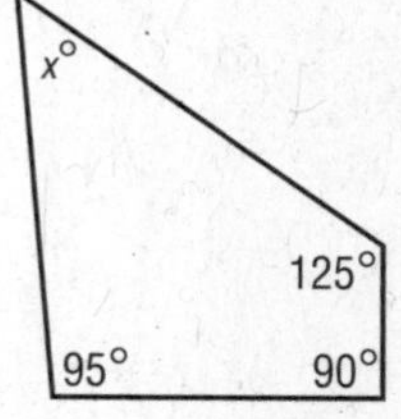

Multi-Part Lesson 3 PART B

NAME ______________________ DATE ____________ PERIOD ______

Skills Practice

Properties of Quadrilaterals

Classify each quadrilateral.

1. 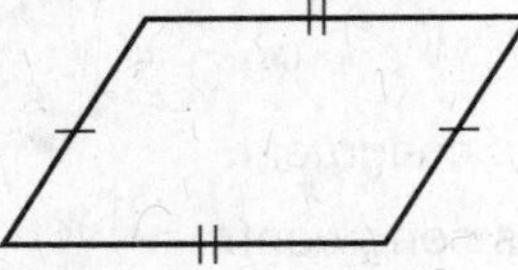

2. 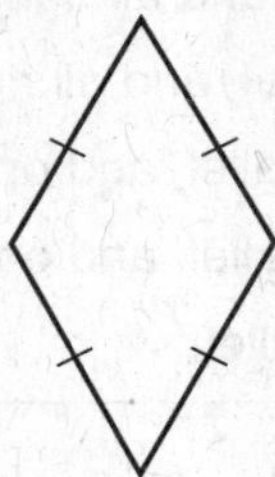

3. 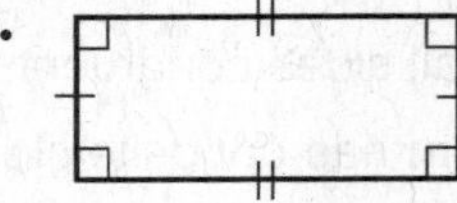

Find the value of x in each quadrilateral.

4. 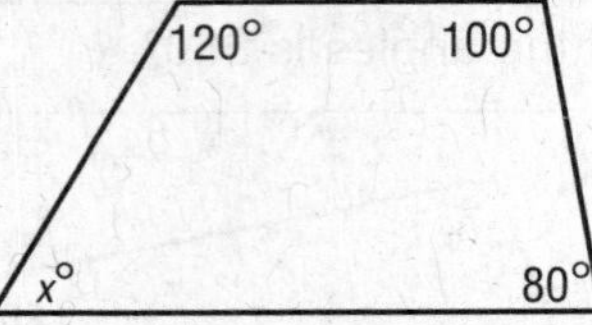

5. 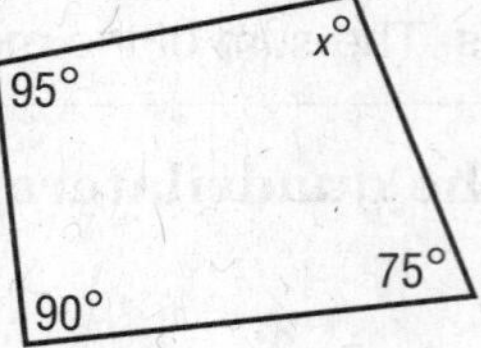

6. 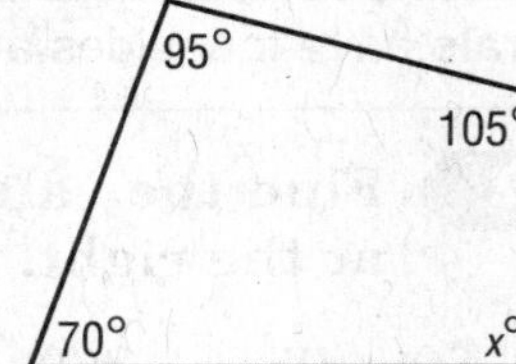

7. 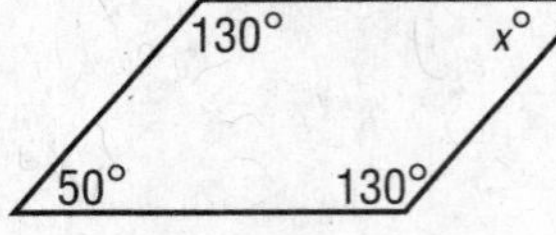

8. 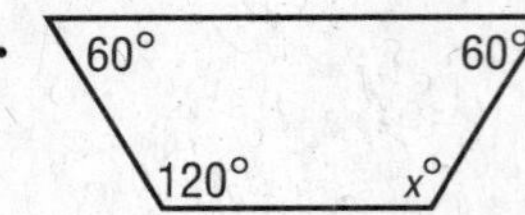

9. 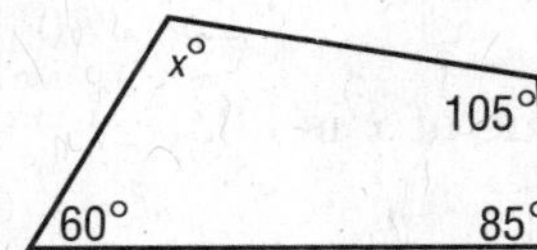

Classify each polygon. Then describe in what ways the figures are the same and in what ways they are different.

10. 

11. 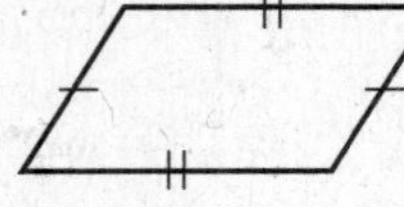

NAME ______________________ DATE ____________ PERIOD _____

Reteach

Properties of Polygons

A **polygon** is a simple, closed figure formed by three or more straight lines. A simple figure does not have lines that cross each other. You have drawn a closed figure when your pencil ends up where it started. Polygons can be classified by the number of sides they have.

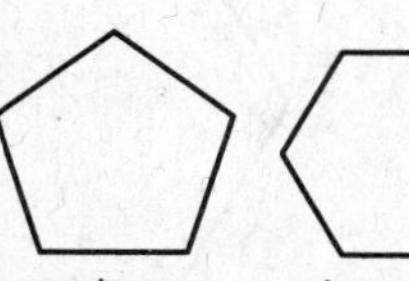
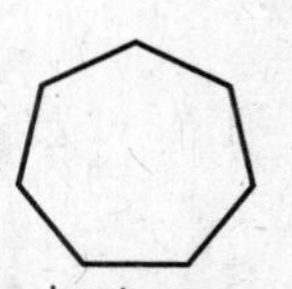
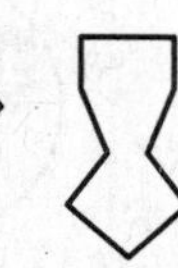
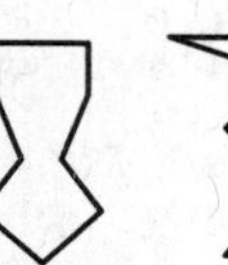

pentagon 5 sides | hexagon 6 sides | heptagon 7 sides | octagon 8 sides | nonagon 9 sides | decagon 10 sides

A polygon that has all sides congruent and all angles congruent is called a **regular polygon**.

Examples **Determine whether each figure is a polygon. If it is, classify the polygon. If it is *not* a polygon, explain why.**

1

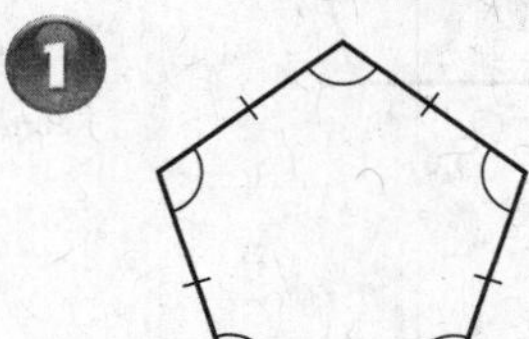

The figure is a regular pentagon.

2

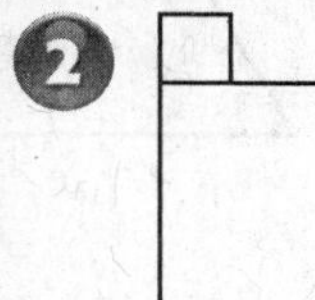

The figure is not a polygon because it has sides that overlap.

Exercises

Determine whether each figure is a polygon. If it is, classify the polygon. If it is *not* a polygon, explain why.

1.

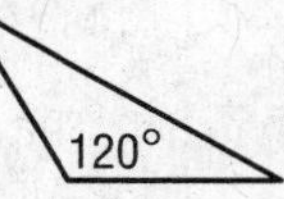

2.

3.

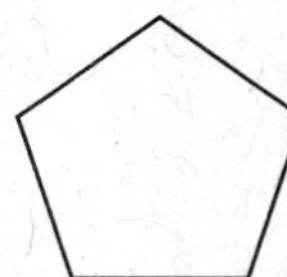

4.

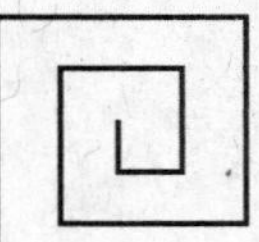

5.

6.

Multi-Part Lesson 3 PART C

NAME ______________________ DATE ____________ PERIOD _____

Skills Practice

Properties of Polygons

Determine whether each figure is a polygon. If it is, classify the polygon. If it is *not* a polygon, explain why.

1.

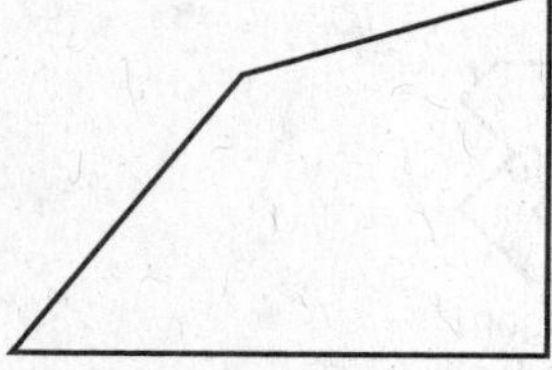

2.

3.

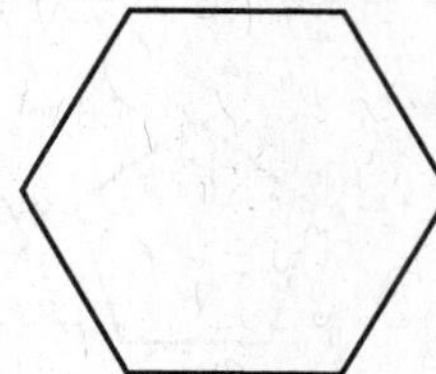

4.

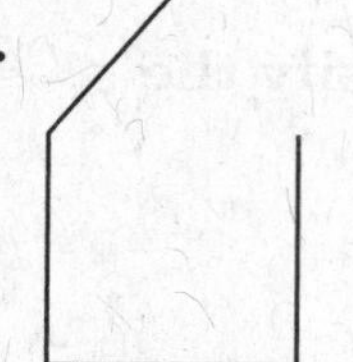

5.

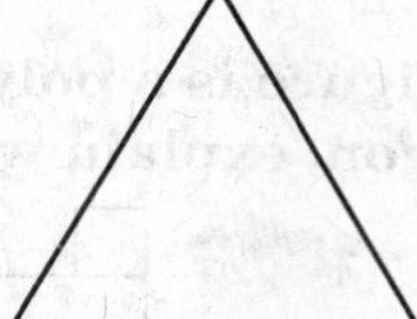

6.

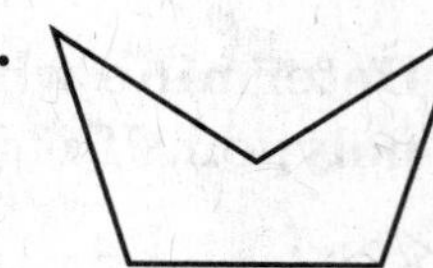

7.

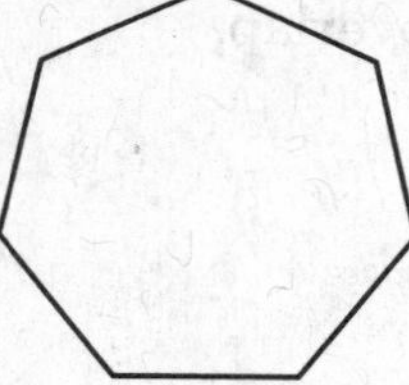

8.

9.

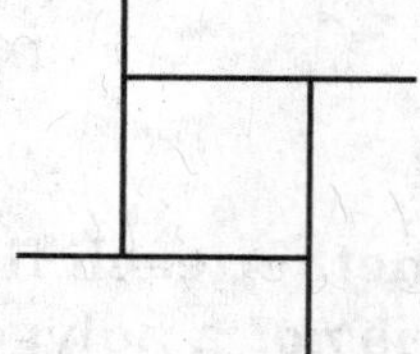

Find the sum of the angle measures of each polygon.

10. nonagon

11. 18-sided polygon

12. 24-sided polygon

Classify the polygons that are used to create the patterns.

13.

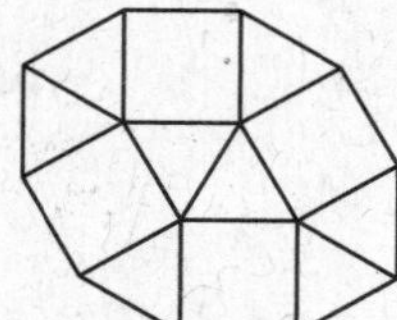

14.

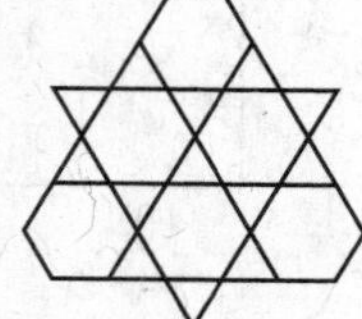

15. What is the perimeter of a regular pentagon with sides 8.4 inches long?

NAME ______________________ DATE ____________ PERIOD ______

Reteach

Similar and Congruent Figures

Figures that have the same size and shape are **congruent figures**.

Figures that have the same shape but not necessarily the same size are **similar figures**.

Examples **Tell whether each pair of figures is *similar*, *congruent*, or *neither*.**

1

The parallelograms have the same shape but not the same size, so they are similar.

2

The triangles have the same shape and size, so they are congruent.

3

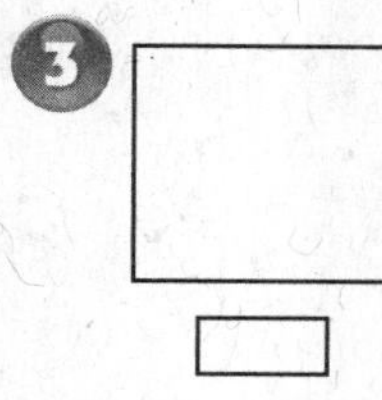

The rectangles are neither the same size nor the same shape, so they are neither congruent nor similar.

Example 4 **The rectangles at the right are similar. What side of rectangle *ABCD* corresponds to side $\overline{ZY}$?**

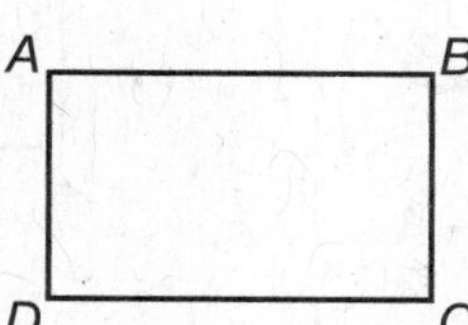

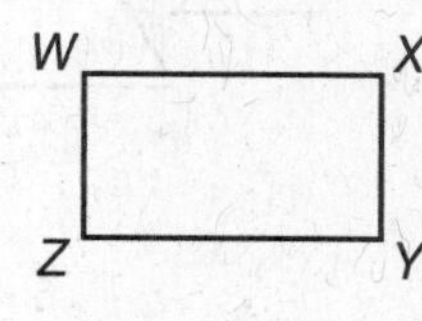

Corresponding sides represent the same side of similar figures. So, side $\overline{DC}$ corresponds to side $\overline{ZY}$.

Exercises

Tell whether each pair of figures is *congruent*, *similar*, or *neither*.

1.

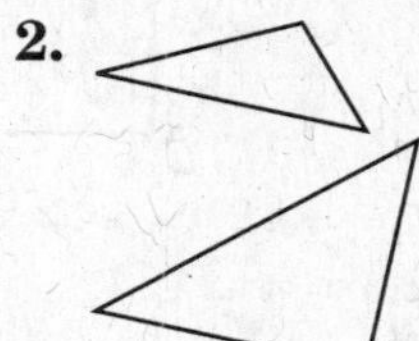

2.

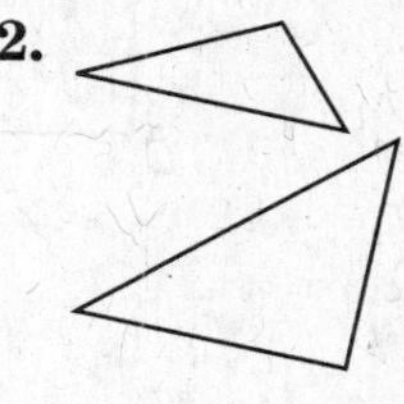

3.

For Exercises 4 and 5, refer to the similar parallelograms at the right.

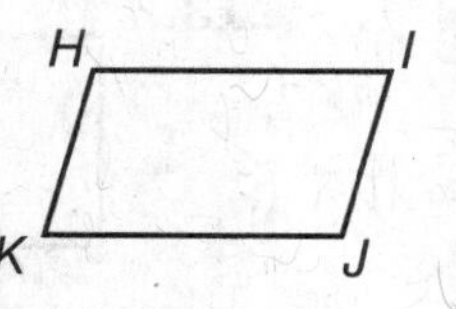

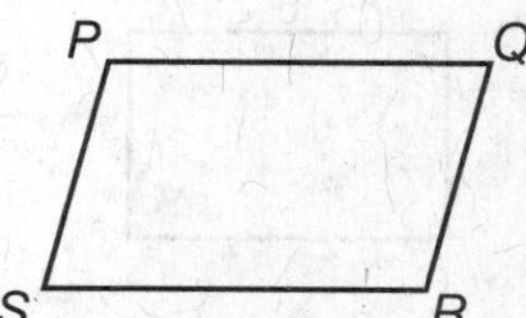

4. What side of parallelogram HIJK corresponds to side $\overline{QR}$?

5. What side or parallelogram PQRS corresponds to side $\overline{HK}$?

NAME ______________________ DATE __________ PERIOD _____

Skills Practice

Similar and Congruent Figures

Tell whether each pair of figures is *similar, congruent,* or *neither.*

1. 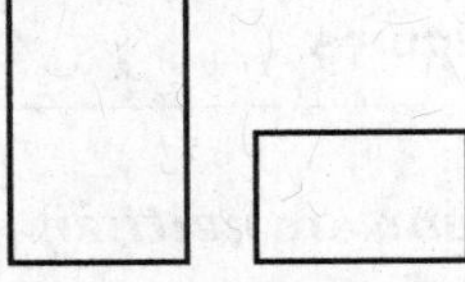

2. 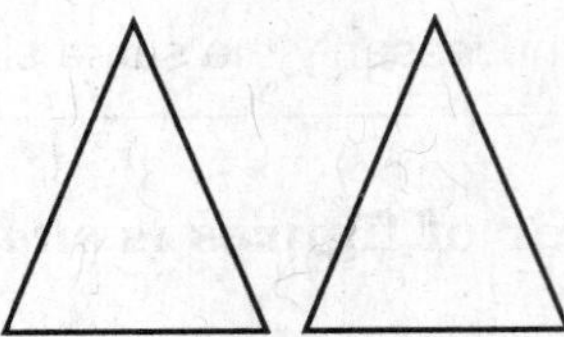

3. 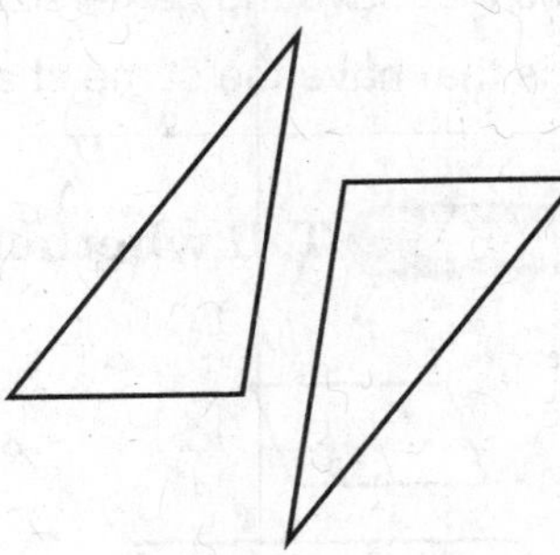

4. 

5. 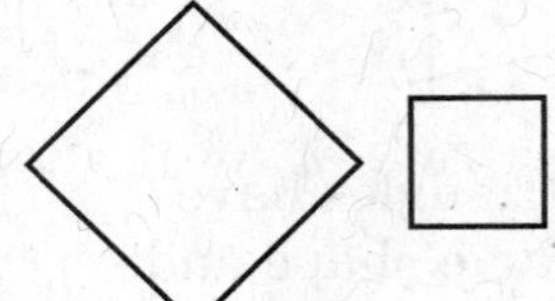

6. 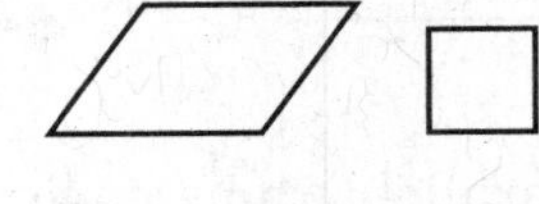

7. 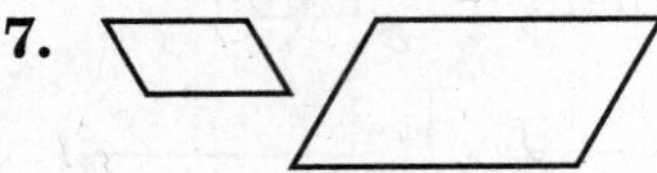

8. 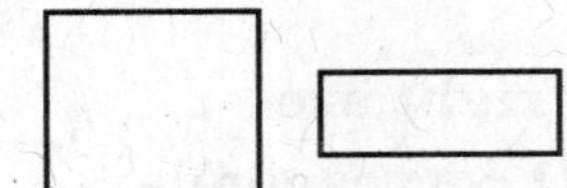

9. 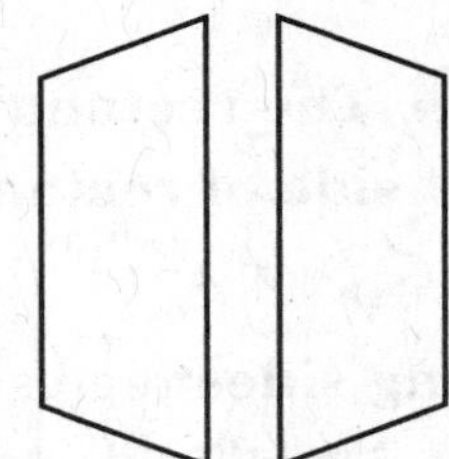

For Exercises 10–14, refer to the similar rectangles at the right.

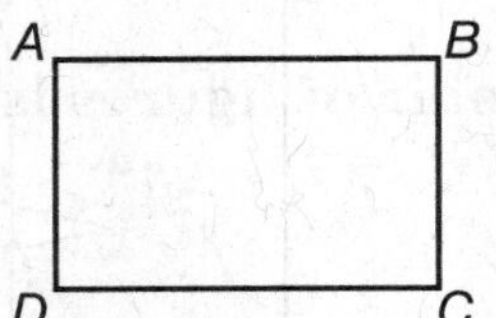

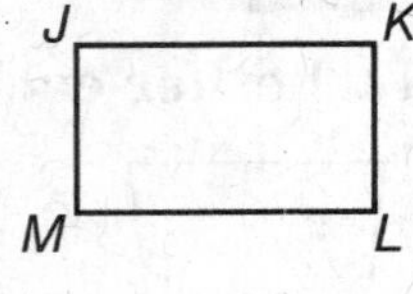

10. What side of rectangle *ABCD* corresponds to side $\overline{JK}$?

11. What side or rectangle *JKLM* corresponds to side $\overline{BC}$?

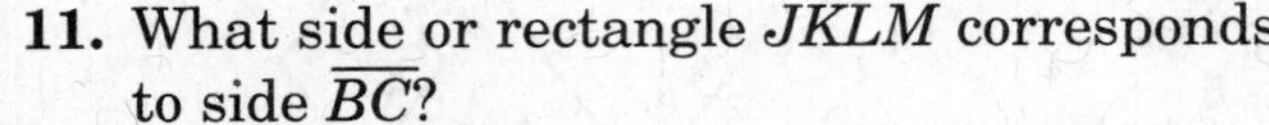

State whether each rectangle is similar to rectangle *ABCD*.

12. 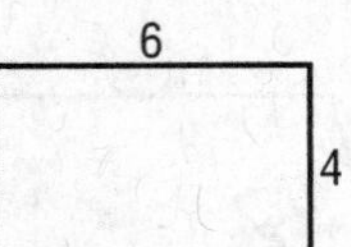

13. 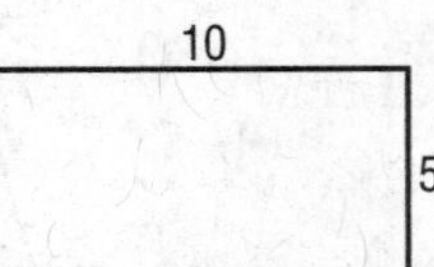

14. 9, 6

Multi-Part Lesson 4 PART B

NAME ______________________ DATE ____________ PERIOD _____

Reteach

Translations

- A transformation is a movement of a geometric figure.
- The resulting figure is called the image.
- A translation is the sliding of a figure without turning it.
- A translation does not change the size or shape of a figure.

Example 1 **Translate triangle *ABC* 5 units to the right.**

Step 1 Move each vertex of the triangle 5 units right. Label the new vertices *A'*, *B'*, *C'*.

Step 2 Connect the new vertices to draw the triangle. The coordinates of the vertices of the new triangle are *A'*(2, 4), *B'*(2, 2), and *C'*(5, 0).

Step 1

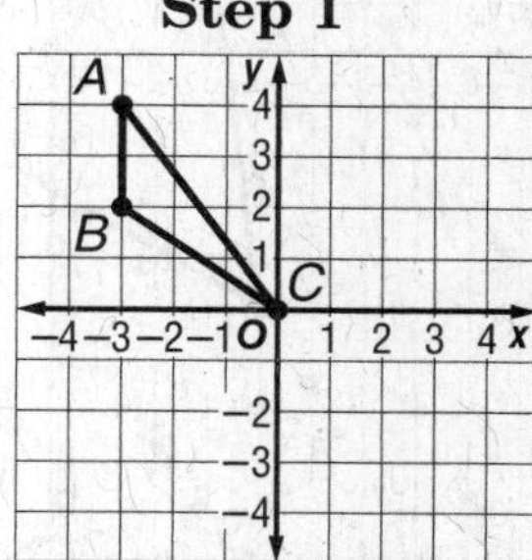

Example 2 **A placemat on a table has vertices at (0, 0), (3, 0), (3, 4), and (0, 4). Find the vertices of the placemat after a translation of 4 units right and 2 units up.**

Step 2

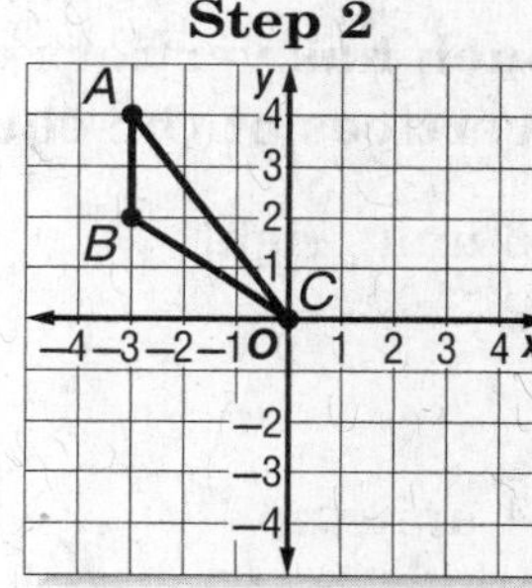

Vertex	$(x + 4, y + 2)$	New vertex
(0, 0)	(0 + 4, 0 + 2)	(4, 2)
(3, 0)	(3 + 4, 0 + 2)	(7, 2)
(3, 4)	(3 + 4, 4 + 2)	(7, 6)
(0, 4)	(0 + 4, 4 + 2)	(4, 6)

Exercises

Find the coordinates of the image of (2, 4), (1, 5), (1, –3), and (3, –4) under each transformation.

1. 2 units right

2. 4 units down

3. 3 units left and 4 units down

4. 5 units right and 3 units up

NAME ______________________ DATE __________ PERIOD _____

Skills Practice

Translations

1. Translate $ABCD$ 5 units down. Graph $A'B'C'D'$.

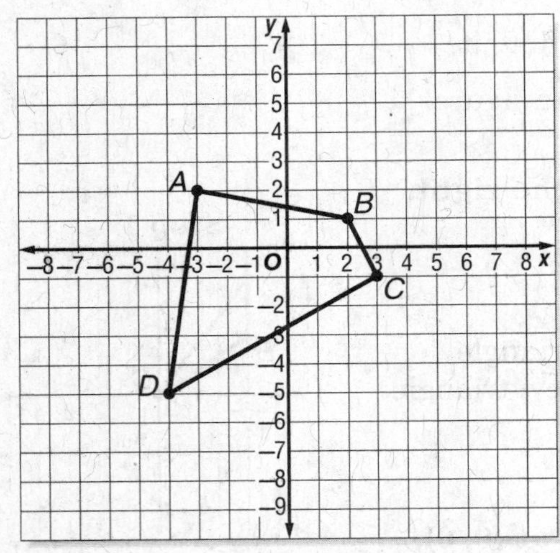

2. Translate PQR 2 units left and 3 units up. Graph $P'Q'R'$.

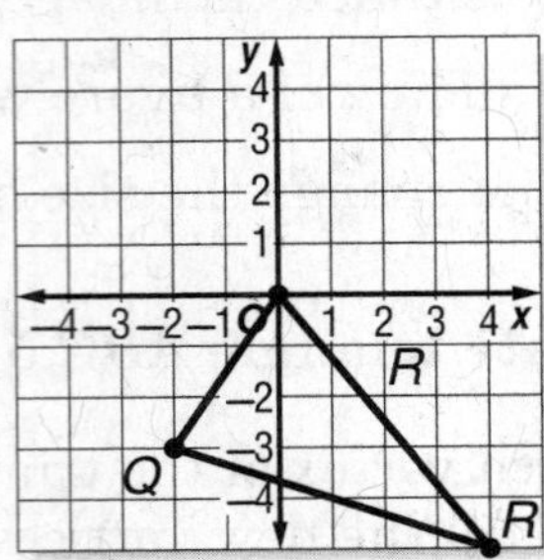

A couch has vertices of (0, 0), (5, 0), (5, 4), and (0, 4) on a floor. Find the vertices of the couch after each translation.

3. 6 units right

4. 5 units left

5. 4 units up

6. 6 units down

7. 2 units left and 3 units up

8. 7 units right and 8 units up

9. 4 units left and 6 units down

10. 8 units right and 2 units down

11. 9 units left and 1 unit up

12. 3 units right and 10 units down

13. 5 units left and 7 units down

14. 3 units right and 3 units up

NAME ______________________ DATE ____________ PERIOD _____

Reteach

Reflections

- A reflection is the mirror image that is created when a figure is flipped over a line.
- A reflection is a type of geometric transformation.
- When reflecting over the x-axis, the y-coordinate changes to its opposite.
- When reflecting over the y-axis, the x-coordinate changes to its opposite.

Example 1 **Reflect triangle ABC over the x-axis.**

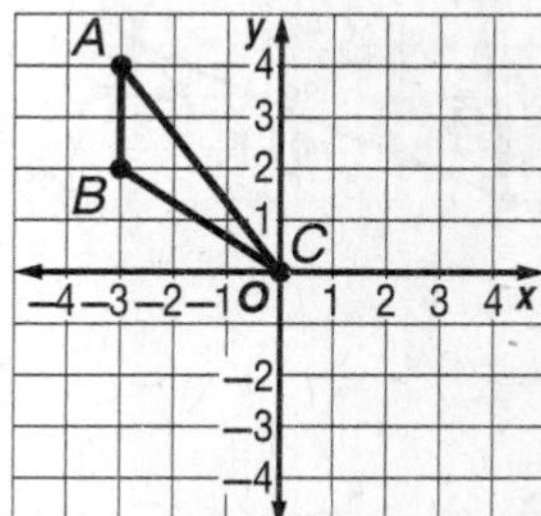

Step 1 Graph triangle ABC on a coordinate plane. Then count the number of units between each vertex and the x-axis.

A is 4 units from the axis.
B is 2 units from the axis.
C is 0 units from the axis.

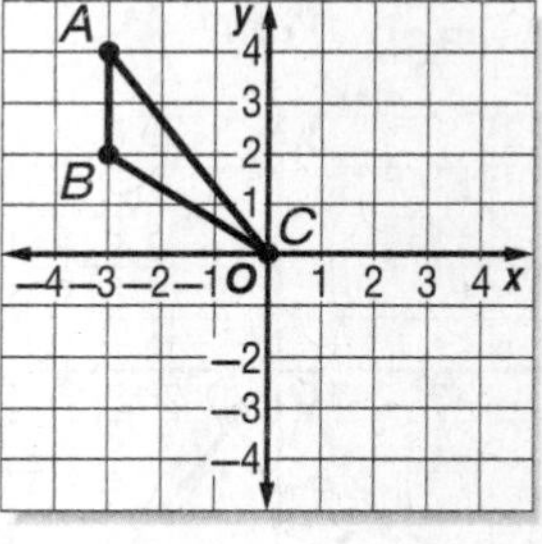

Step 2 Make a point for each vertex the same distance away from the x-axis on the opposite side of the x-axis and connect the new points to form the image of the triangle. The new points are $A'(-3, -4)$, $B'(-2, -2)$, and $C'(0, 0)$.

Exercises

Find the coordinates of the image of (2, 4), (1, 5), (1, –3) and (3, –4) under each transformation.

1. a reflection over the x-axis

2. a reflection over the y-axis.

Find the coordinates of the image of (–1, 1), (3, –2) and (0, 5) under each transformation.

3. a reflection over the x-axis

4. a reflection over the y-axis

NAME ______________________ DATE __________ PERIOD _____

Skills Practice

Reflections

1. Reflect *ABCD* over the *x*-axis. Graph *A′B′C′D′*.

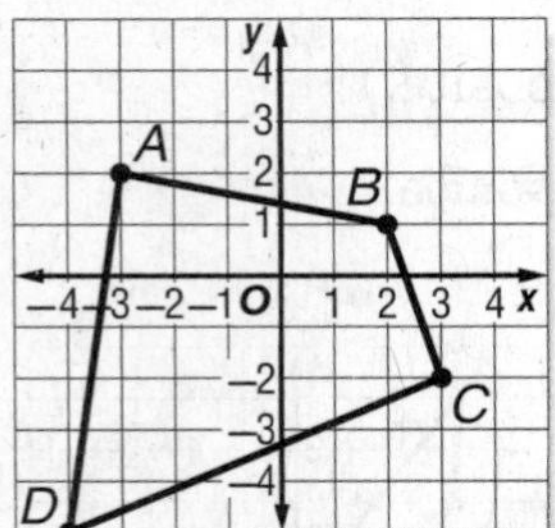

2. Reflect *ABCD* over the *y*-axis. Graph *A′B′C′D′*.

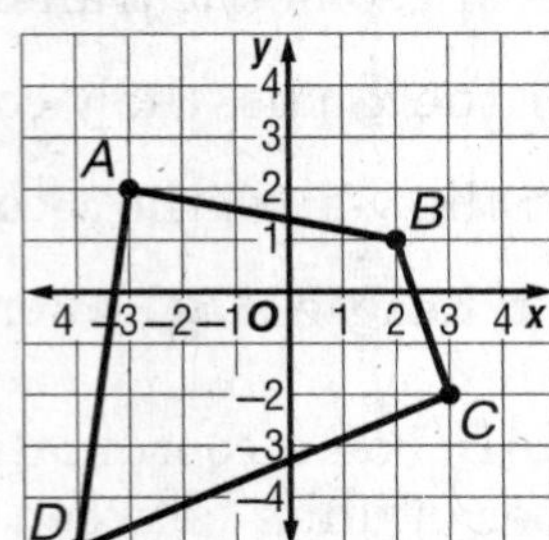

3. Reflect *PQR* over the *x*-axis. Graph *P′Q′R′*.

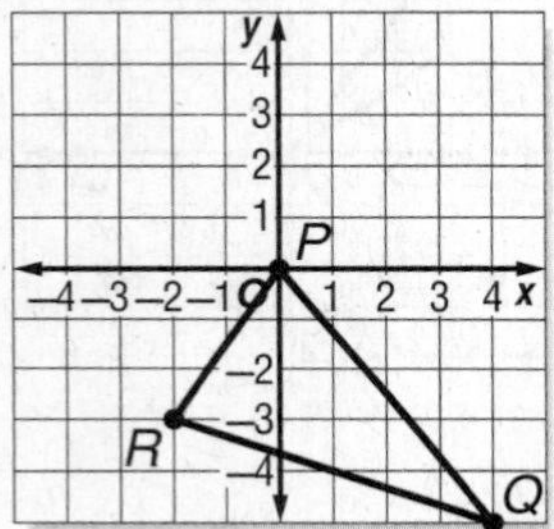

4. Reflect *PQR* over the *y*-axis. Graph *P′Q′R′*.

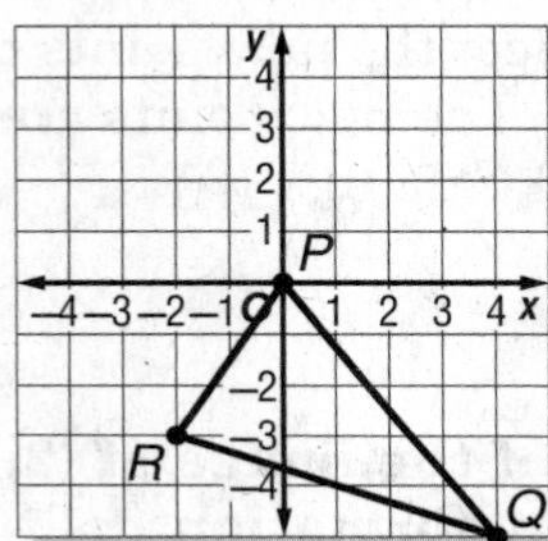

A couch has vertices of (0, 0), (5, 0), (5, 4), and (0, 4) on a floor. Find the vertices of the couch after each transformation.

5. a reflection over the *x*-axis

6. a reflection over the *y*-axis.

A rug has vertices of (4, 1), (2, 5), and (−3, −2) on a floor. Find the vertices of the rug after each transformation.

7. a reflection over the *x*-axis

8. a reflection over the *y*-axis

NAME ______________________ DATE __________ PERIOD ____

Reteach

Rotations

- A rotation occurs when a figure is rotated around a point.
- Another name for a rotation is a turn.
- In a rotation clockwise of 90° about the origin, the point (x, y) becomes $(y, -x)$.
- In a rotation clockwise of 180° about the origin, the point (x, y) becomes $(-x, -y)$.
- In a rotation clockwise of 270° about the origin, the point (x, y) becomes $(-y, x)$.

Example 1 **Rotate triangle *ABC* clockwise 180° about the origin.**

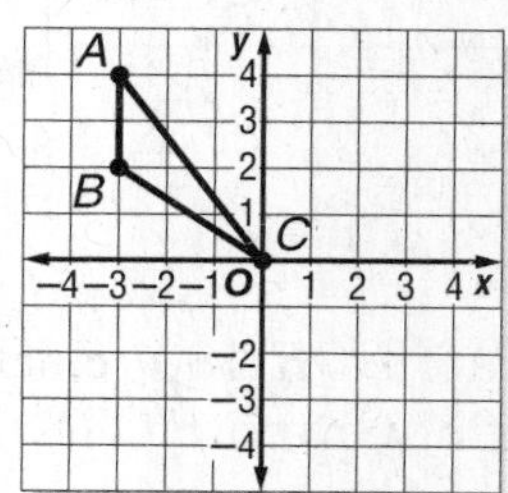

Step 1 Graph triangle *ABC* on a coordinate plane.

Step 2 Sketch segment *AO* connecting point *A* to the origin. Sketch another segment *A′ O* so that the angle between point *A*, *O*, and *A′* measures 180° and the segment is congruent to *AO*.

Step 3 Repeat for point *B* (point *C* won't move since it is at the origin). Then connect the vertices to form triangle *A′B′C′*.

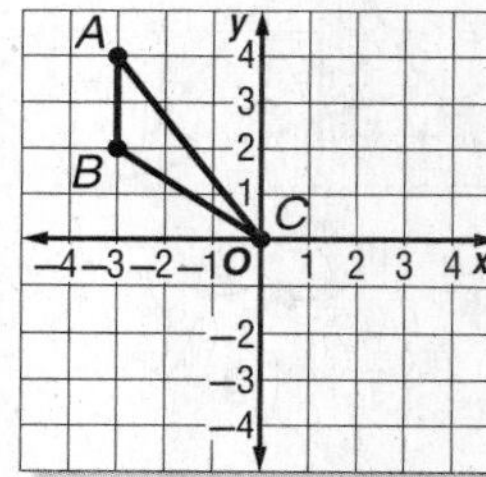

Exercises

Find the coordinates of the image of (2, 4), (1, 5), (1, −3) and (3, −4) under each transformation.

1. a rotation of 90° about the origin

2. a rotation of 270° about the origin

Determine whether each figure has rotational symmetry. Write *yes* or *no*. If yes, name the angle(s) of rotation.

3. X

4. Q

NAME ______________________ DATE __________ PERIOD _____

Multi-Part Lesson 4 PART D

Skills Practice

Rotations

1. Rotate $ABCD$ 90° clockwise about the origin. Graph $A'B'C'D'$.

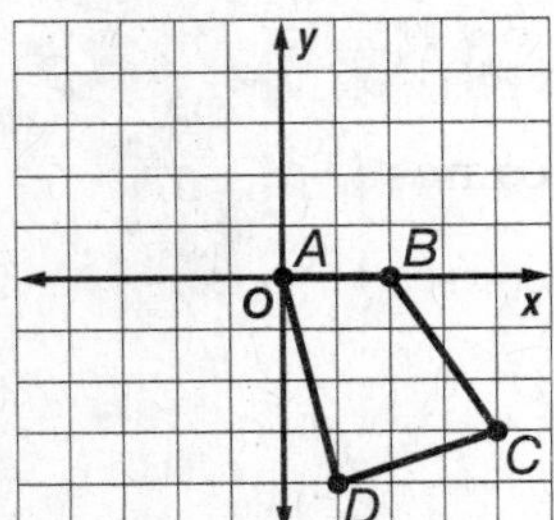

2. Rotate $ABCD$ 270° clockwise about the origin. Graph $A'B'C'D'$.

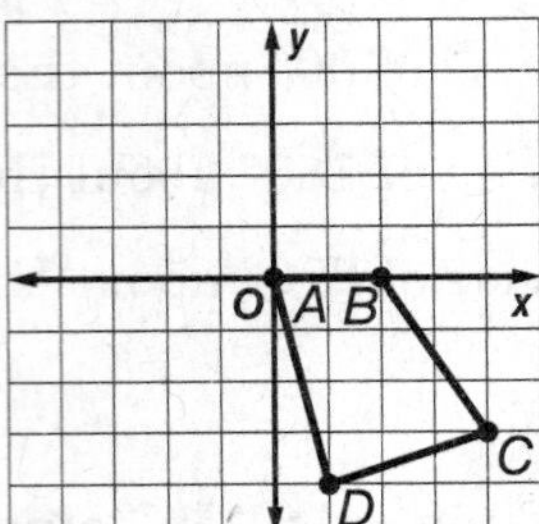

3. Rotate PQR 180° clockwise about the origin. Graph $P'Q'R'$.

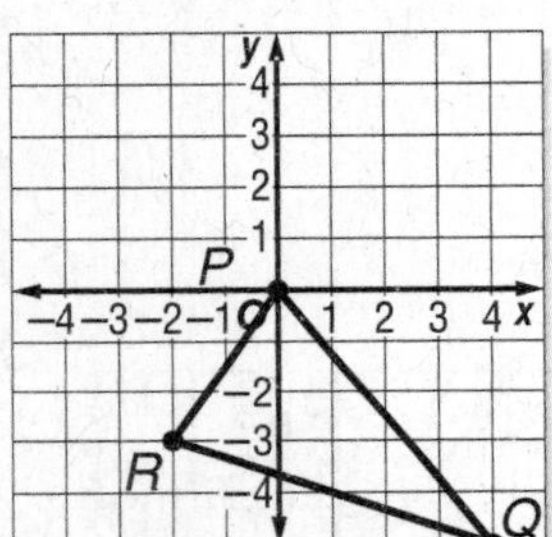

4. Rotate PQR 90° counterclockwise about the origin. Graph $P'Q'R'$.

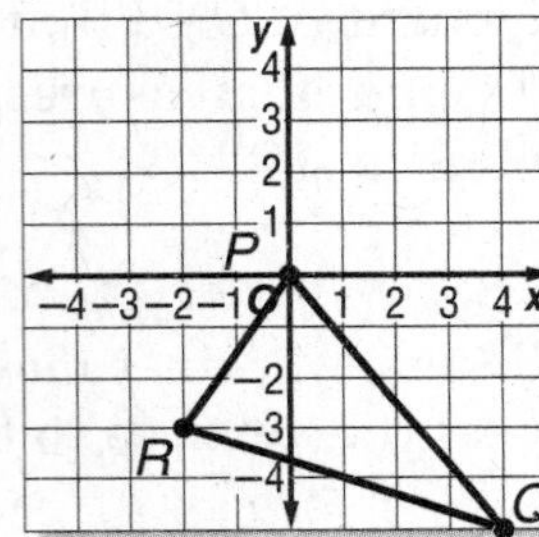

Determine whether each figure has rotational symmetry. Write *yes* or *no*. If yes, name the angle(s) of rotation.

5. A

6. 

7.

8.

NAME ______________________ DATE ____________ PERIOD ______

Reteach

Circumference

center

The **circumference,** C, is the distance around a circle.

The **diameter,** d, is the distance across a circle through its center.

The **radius,** r, is the distance from the center to any point on a circle.

The diameter of a circle is twice its radius. $d = 2r$

The radius is half the diameter. $r = \frac{d}{2}$

The circumference of a circle is equal to π times its diameter or π times twice its radius. $C = \pi d$ $C = 2\pi r$

Example 1 **The radius of a circle is 7 meters. Find the diameter.**

$d = 2r$

$d = 2 \cdot 7$ Replace r with 7.

$d = 14$ Multiply.

The diameter is 14 meters.

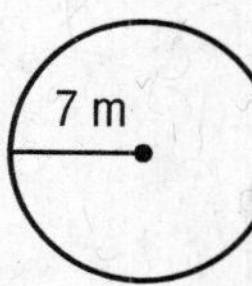

Example 2 **Find the circumference of a circle with a radius that is 13 inches. Use 3.14 for π. Round to the nearest tenth.**

$C = 2\pi r$ Write the formula.

$C \approx 2 \times 3.14 \times 13$ Replace r with 13 and π with 3.14.

$C \approx 81.64$ Multiply.

Rounded to the nearest tenth, the circumference is about 81.6 inches.

Exercises

Find the circumference of each circle. Use 3.14 or $\frac{22}{7}$ for π. Round to the nearest tenth if necessary.

1.

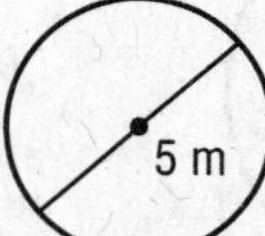

2.

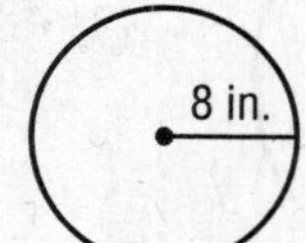

3.

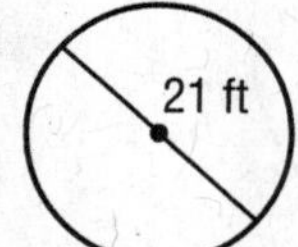

NAME ______________________ DATE ____________ PERIOD ______

Multi-Part Lesson 2 PART B

Skills Practice

Circumference

Find the radius or diameter of each circle with the given dimensions.

1. $r = 13$ cm

2. $d = 4$ ft

3. $r = 10$ mm

4. $d = 16$ in.

5. $r = 7$ mi

6. $d = 22$ yd

Find the circumference of each circle. Use 3.14 or $\frac{22}{7}$ for π. Round to the nearest tenth if necessary.

7.

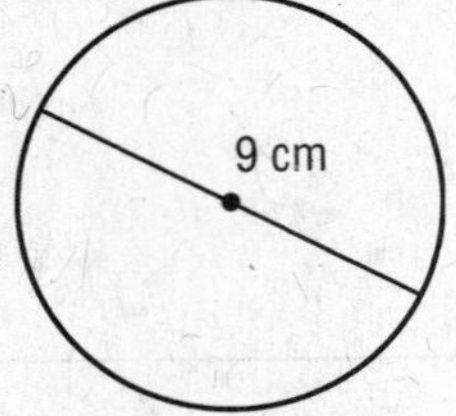

8.

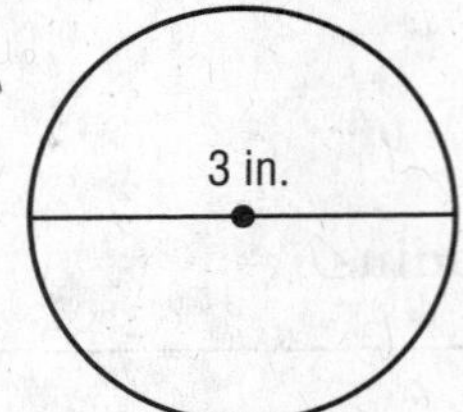

9.

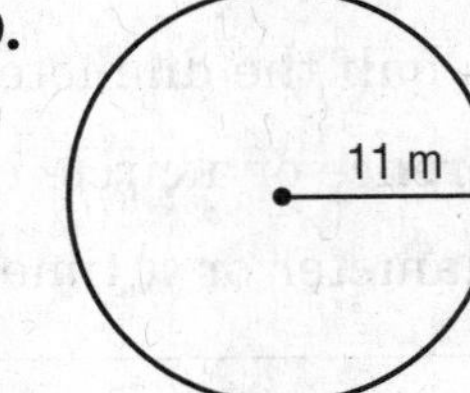

10.

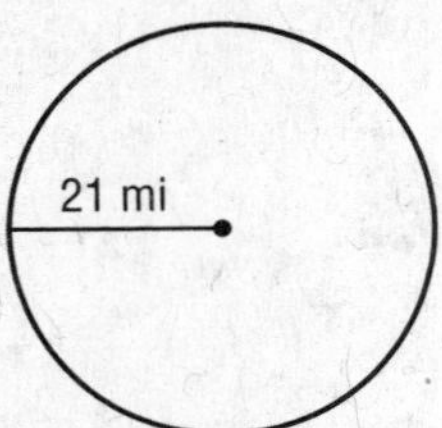

11.

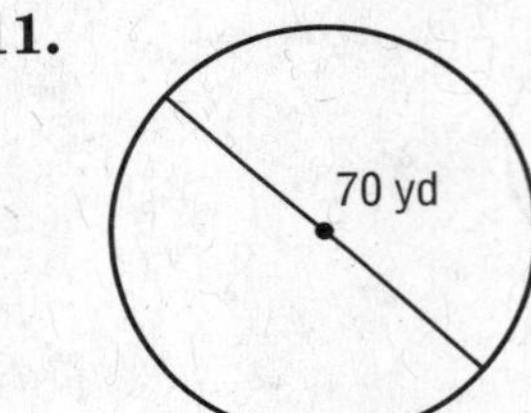

12.

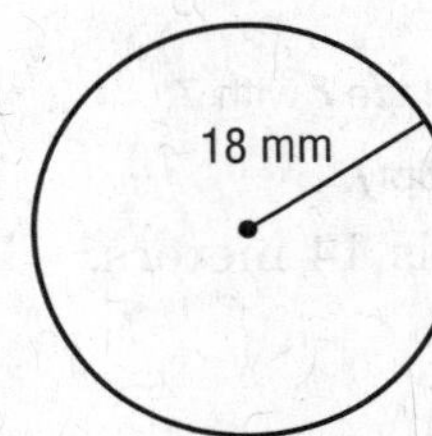

13.

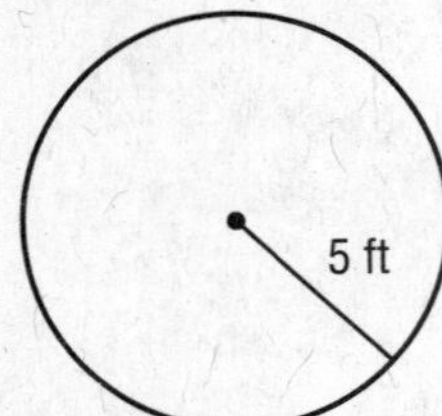

14.

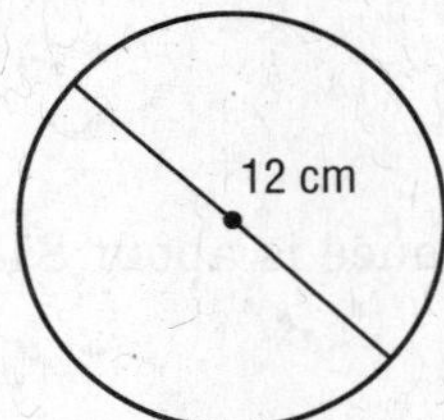

15.

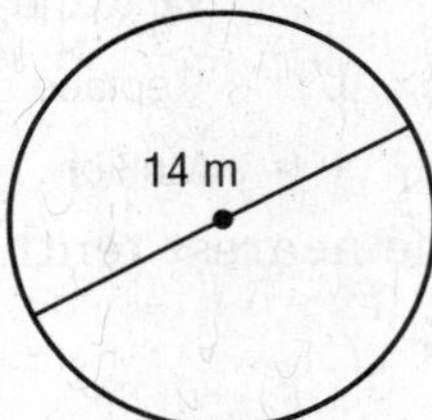

16.

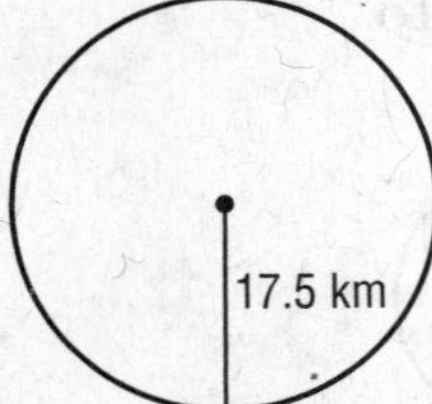

17.

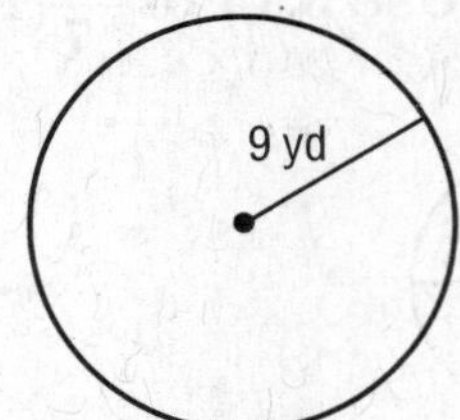

18.

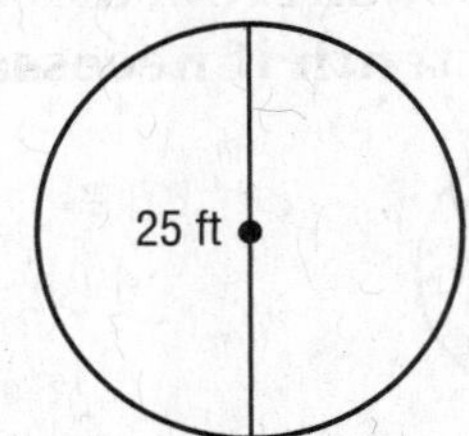

NAME ______________________ DATE __________ PERIOD ______

Reteach

Area of Circles

The area A of a circle equals the product of pi (π) and the square of its radius r.

$A = \pi r^2$.

Example 1 **Find the area of the circle. Use 3.14 for π.**

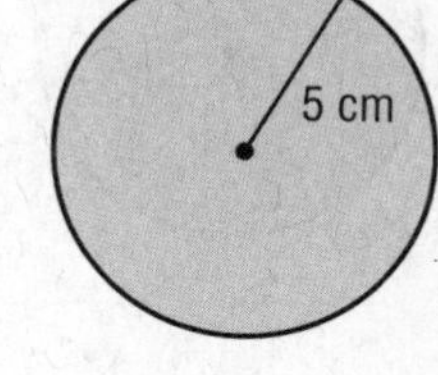

$A = \pi r^2$	Area of circle
$A \approx 3.14 \cdot 5^2$	Replace π with 3.14 and r with 5.
$A \approx 3.14 \cdot 25$	$5^2 = 5 \cdot 5 = 25$
$A \approx 78.5$	

The area of the circle is approximately 78.5 square centimeters.

Example 2 **Find the area of a circle that has a diameter of 9.4 millimeters. Use 3.14 for π. Round to the nearest tenth.**

$A = \pi r^2$	Area of circle
$A \approx 3.14 \cdot 4.7^2$	Replace π with 3.14 and r with $9.4 \div 2$ or 4.7.
$A \approx 69.4$	Multiply.

The area of the circle is approximately 69.4 square millimeters.

The formula for the area of a semicircle, or half a circle, is $A = \frac{1}{2}\pi r^2$.

Exercises

Find the area of each circle. Use 3.14 or $\frac{22}{7}$ for π. Round to the nearest tenth.

1. 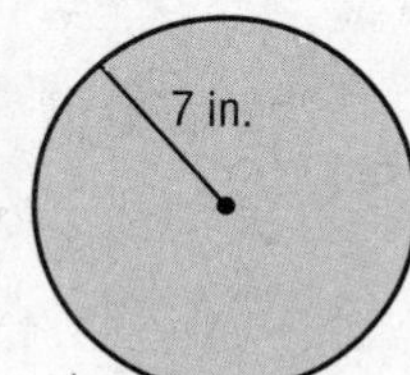

2. 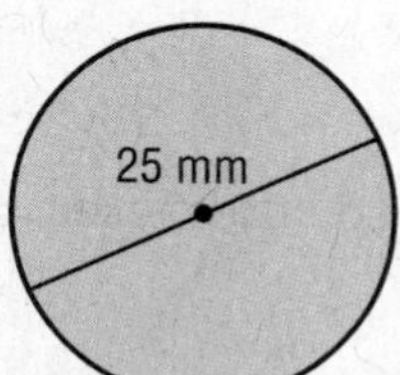

3. 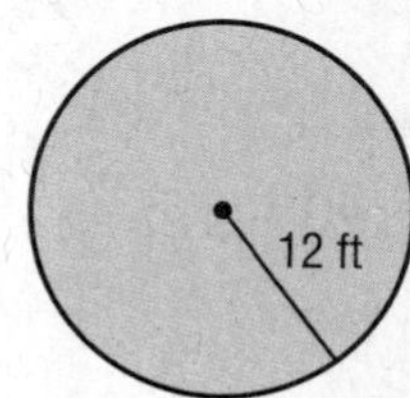

Approximate the area of each semicircle.

4. 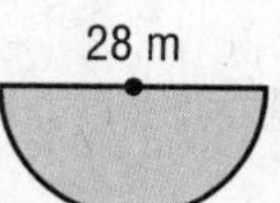

5. 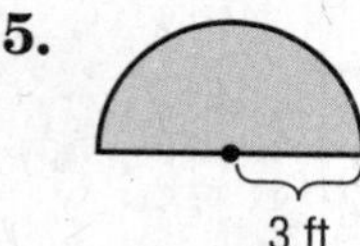

NAME ______________________ DATE ____________ PERIOD _____

Skills Practice

Area of Circles

Find the area of each circle. Round to the nearest tenth. Use 3.14 or $\frac{22}{7}$ for π.

1.

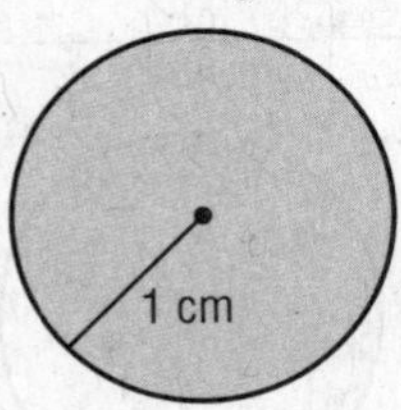

2.

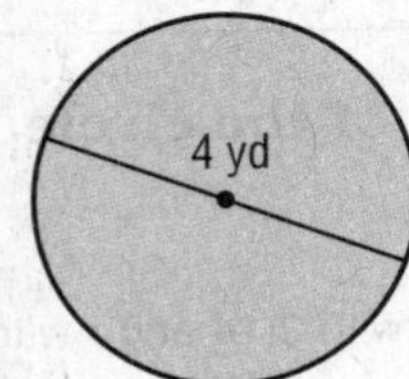

3.

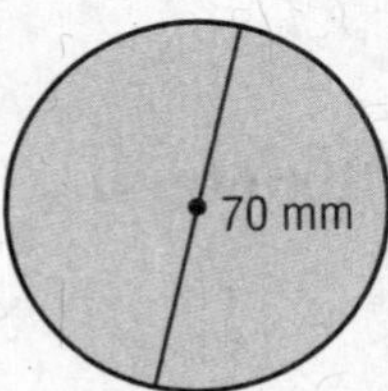

4.

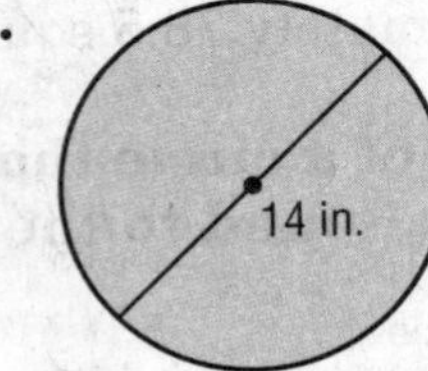

5.

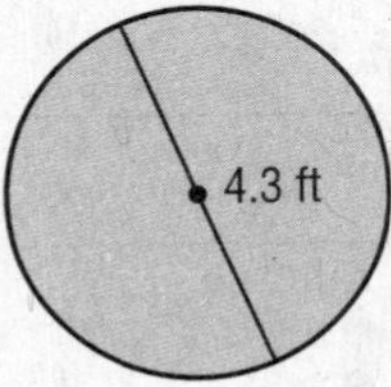

6.

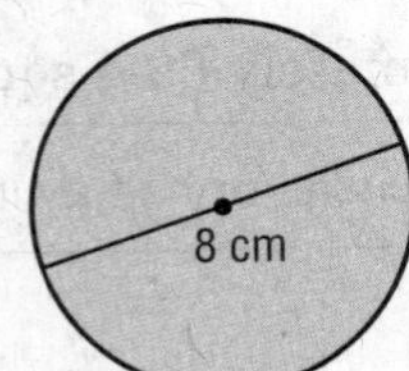

7. radius = 5.7 mm

8. radius = 8.2 ft

9. diameter = 3 in.

10. diameter = 15.6 cm

Find the approximate area of each semicircle.

11.

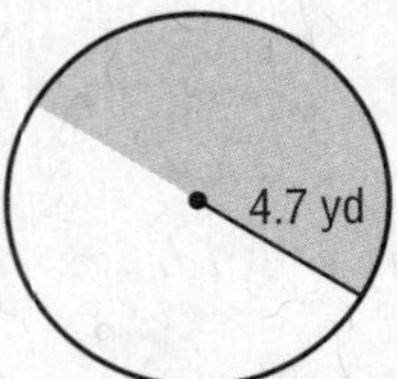

12.

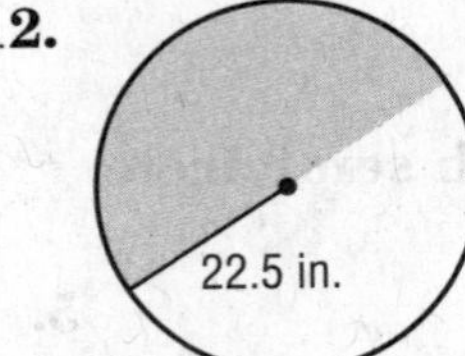

NAME ______________________ DATE ____________ PERIOD _____

Reteach

Perimeter of Composite Figures

The distance around any closed figure is called its **perimeter**. To find the perimeter, add the measures of all the sides of the figure.

Rectangle	The perimeter P of a rectangle is twice the sum of the base b and height h.	$P = b + b + h + h$ $P = 2b + 2h$
Composite Figures	A **composite figure** is made up of triangles, quadrilaterals, semicircles, and other 2-dimensional figures. To find the perimeter of a composite figure, add the distances around the closed figure.	

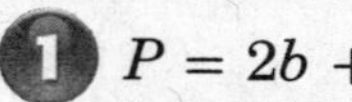

Find the perimeter of each figure.

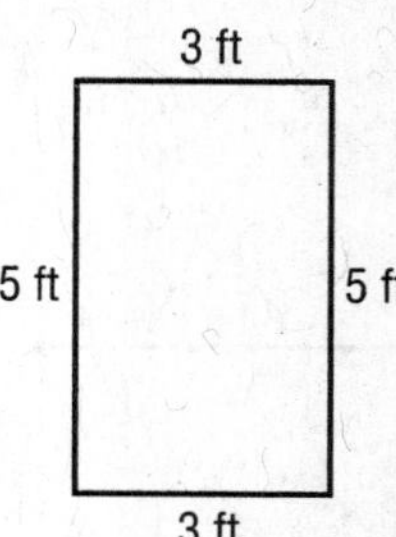

1
$P = 2b + 2h$ Perimeter of a rectangle
$P = 2(3) + 2(5)$ Replace b with 3 and h with 5.
$P = 6 + 10$ Multiply.
$P = 16$ Add.

The perimeter is 16 feet.

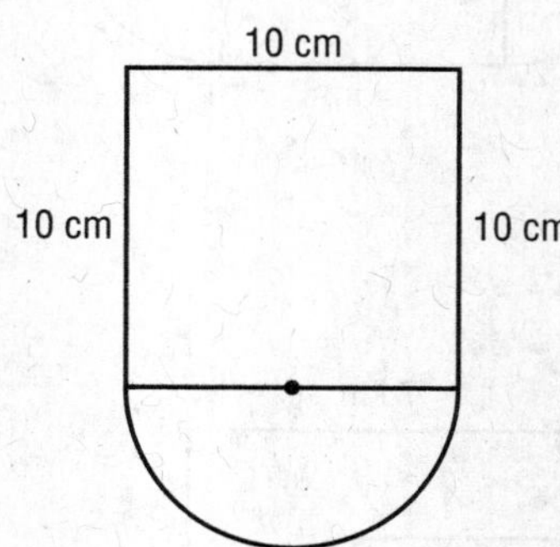

2 Find the circumference of the circle.

$C = \pi d$ Circumference of a circle
$C \approx 3.14(10)$ Replace d with 10.
$C \approx 31.4$ Multiply.

Since you only need half of the circumference, divide by 2.
$31.4 \div 2 = 15.7$
The perimeter is $10 + 10 + 10 + 15.7$ or 45.7 centimeters.

Exercises

Find the perimeter of each figure. Use 3.14 for π.

1.

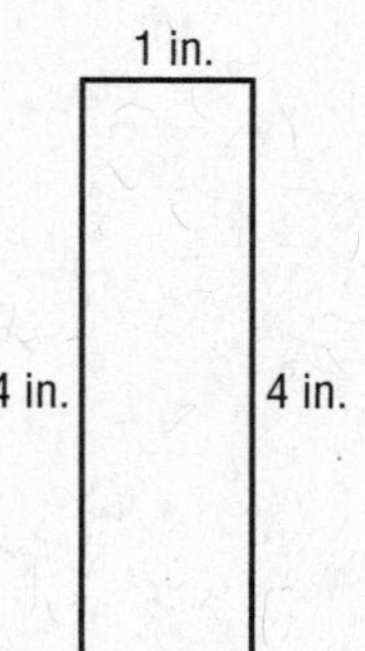

2.

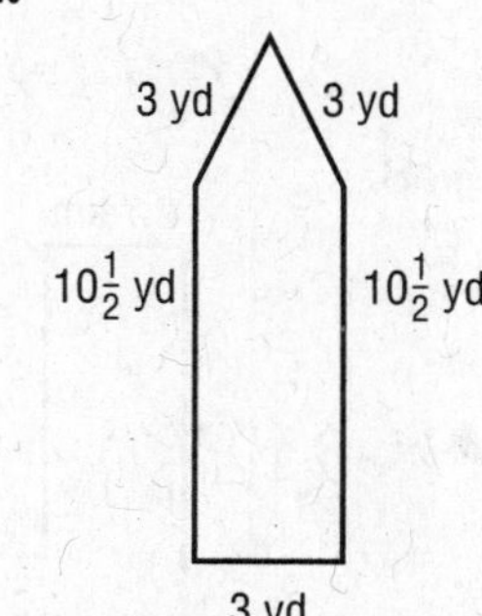

3.

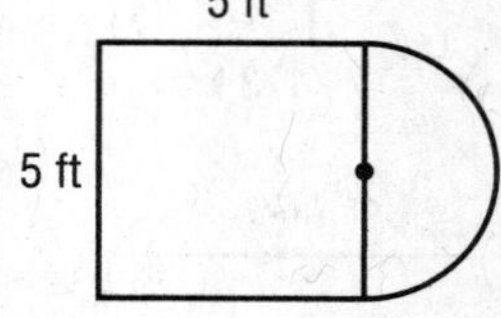

NAME ______________________ DATE ____________ PERIOD _____

Skills Practice

Perimeter of Composite Figures

Find the perimeter of each figure. Use 3.14 for π.

1.

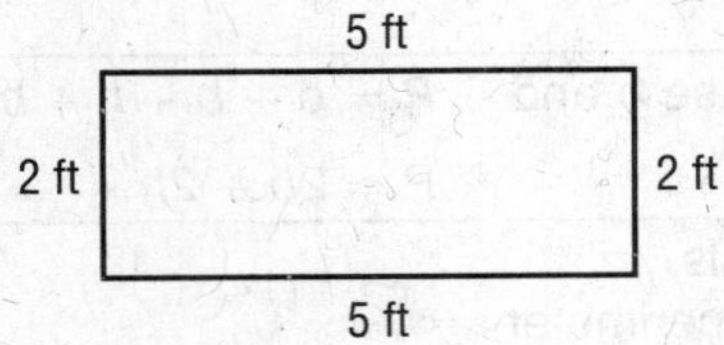

2.

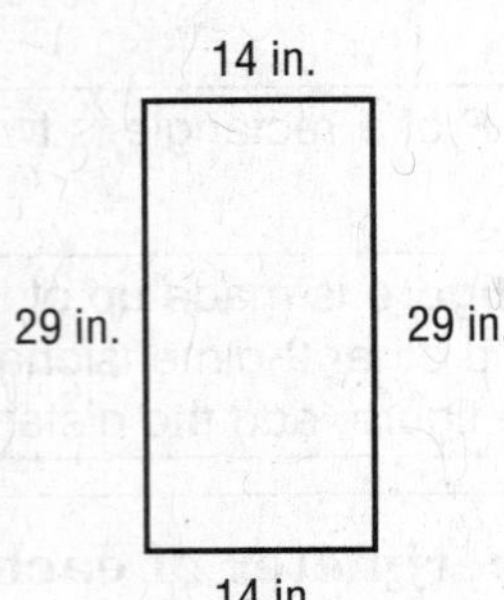

3.

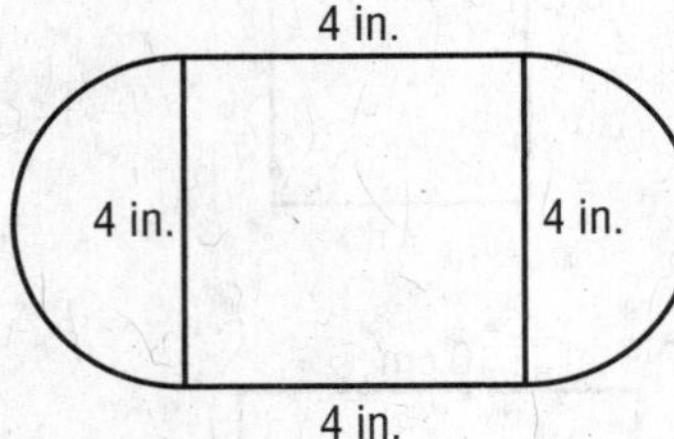

4.

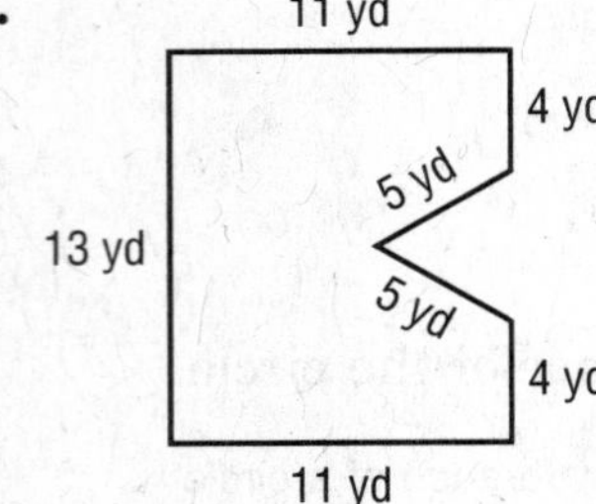

5.

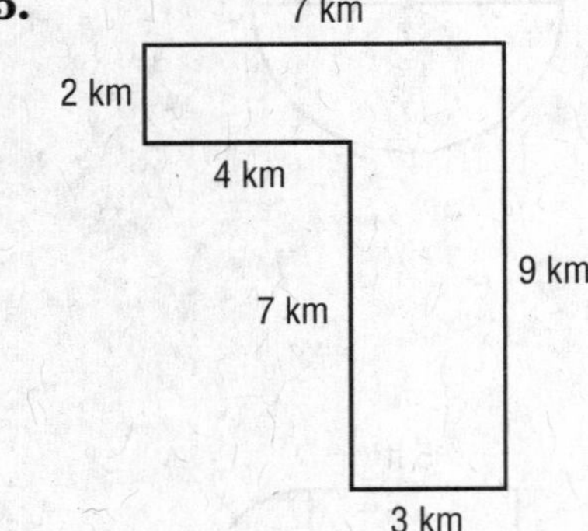

6.

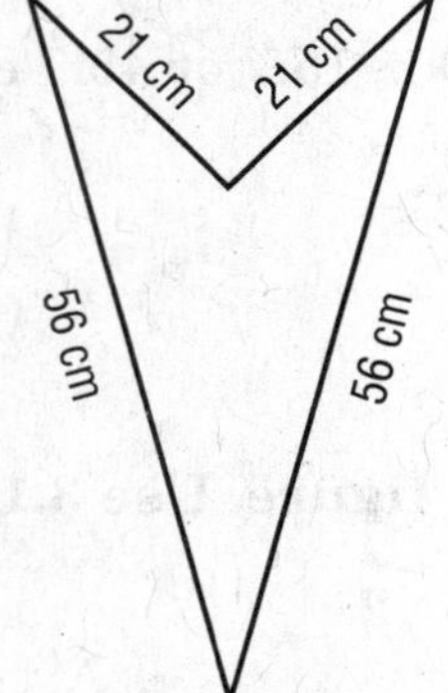

7.

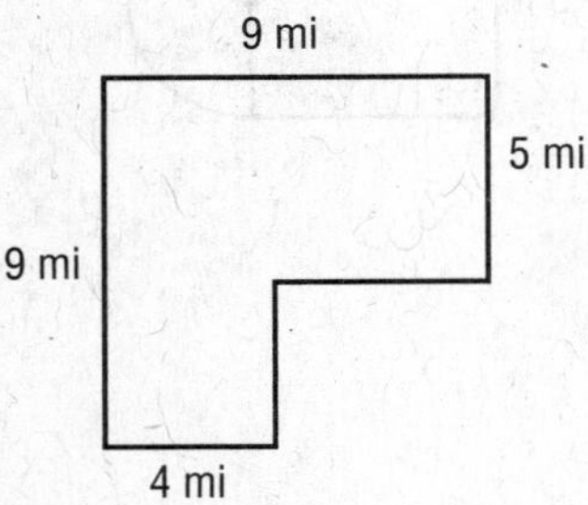

8.

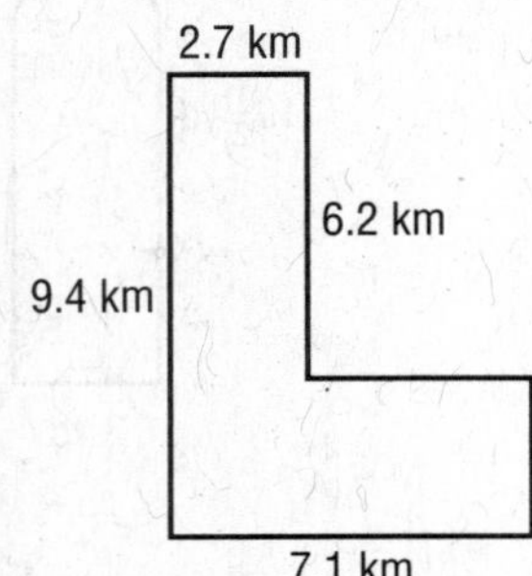

NAME ______________________ DATE ____________ PERIOD _____

Reteach

Area of Composite Figures

To find the area of a composite figure, separate it into figures whose areas you know how to find, and then add the areas.

Example **Find the area of the figure at the right in square feet.**

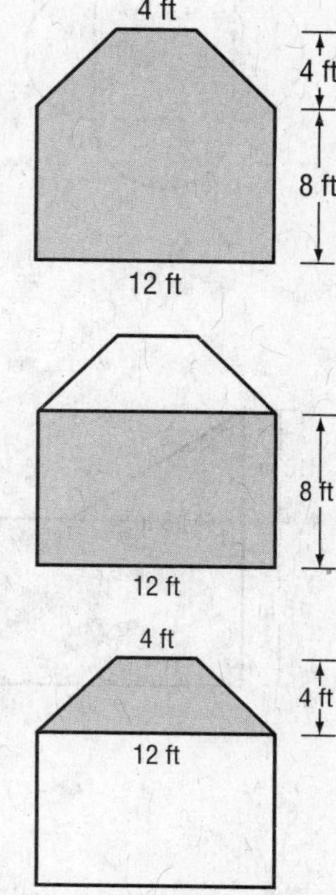

The figure can be separated into a rectangle and a trapezoid. Find the area of each.

Area of Rectangle

$A = \ell w$ Area of a rectangle

$A = 12 \cdot 8$ Replace ℓ with 12 and w with 8.

$A = 96$ Multiply.

Area of Trapezoid

$A = \frac{1}{2}h(b_1 + b_2)$ Area of a trapezoid

$A = \frac{1}{2}(4)(4 + 12)$ Replace h with 4, b_1 with 4, and b_2 with 12.

$A = 32$ Multiply.

The area of the figure is 96 + 32 or 128 square feet.

Exercises

Find the area of each figure. Round to the nearest tenth if necessary.

1.

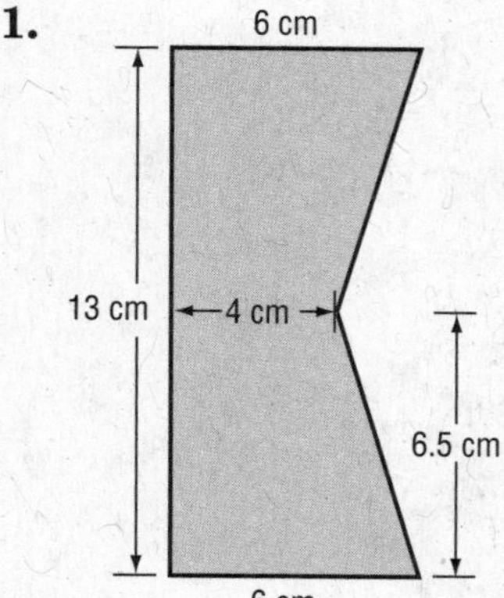

2.

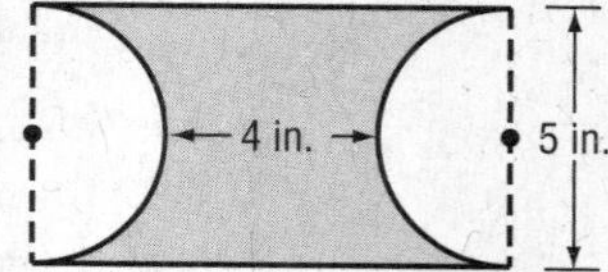

3.

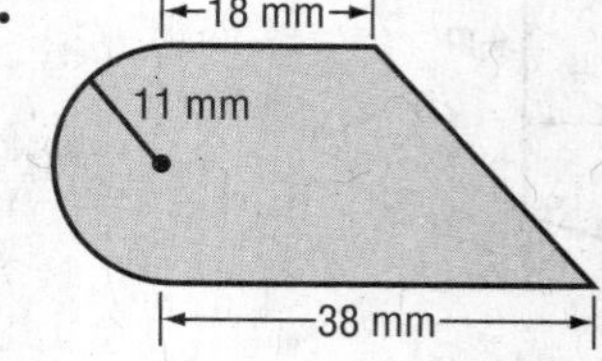

NAME ______________________ DATE ____________ PERIOD _____

Multi-Part Lesson 3 PART C

Skills Practice

Area of Composite Figures

Find the area of each figure. Round to the nearest tenth if necessary.

1.

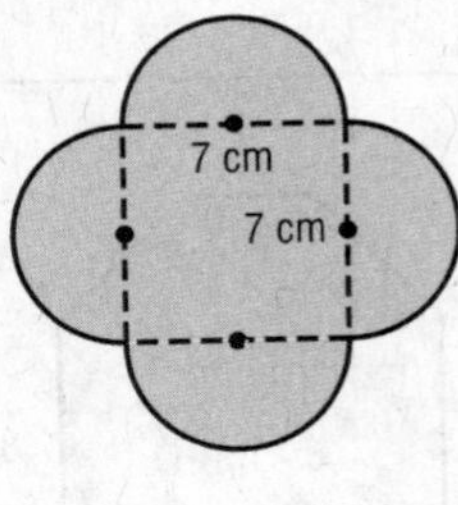

2.

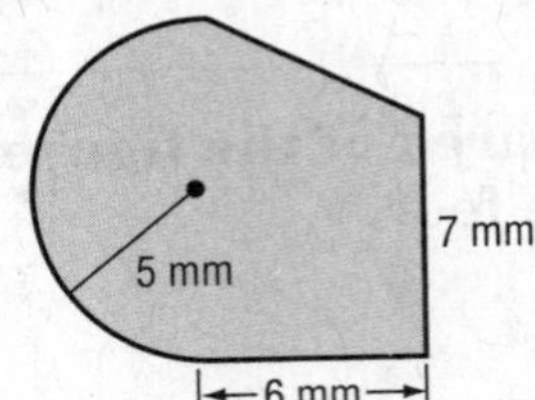

3.

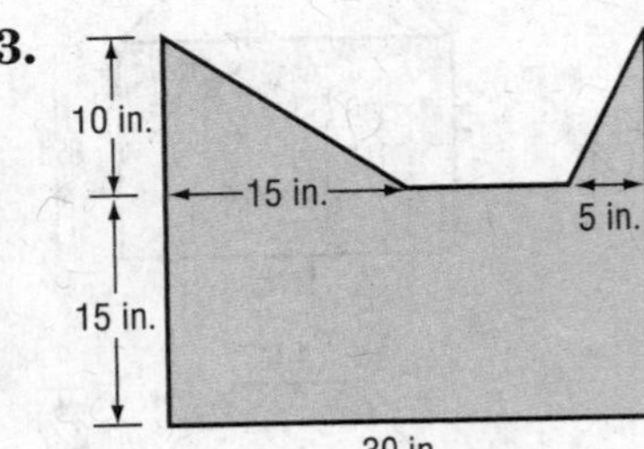

4.

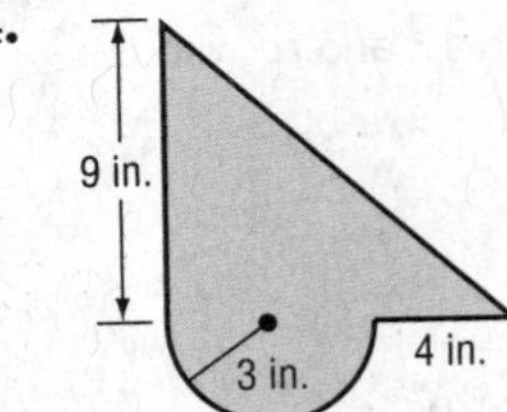

5.

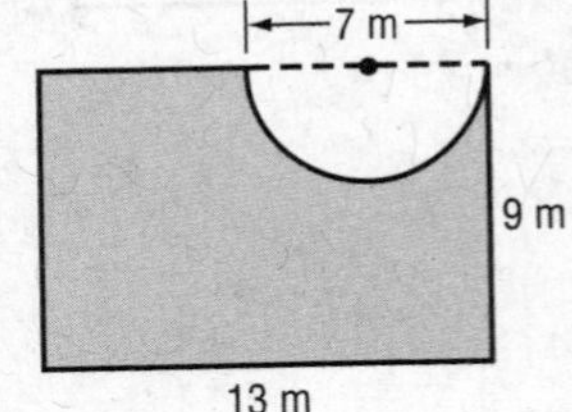

6.

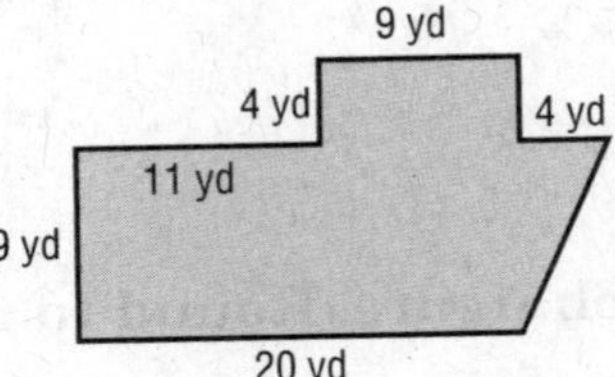

7.

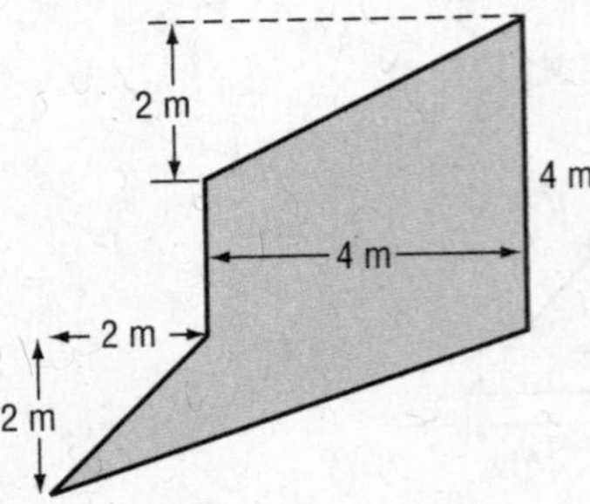

8.

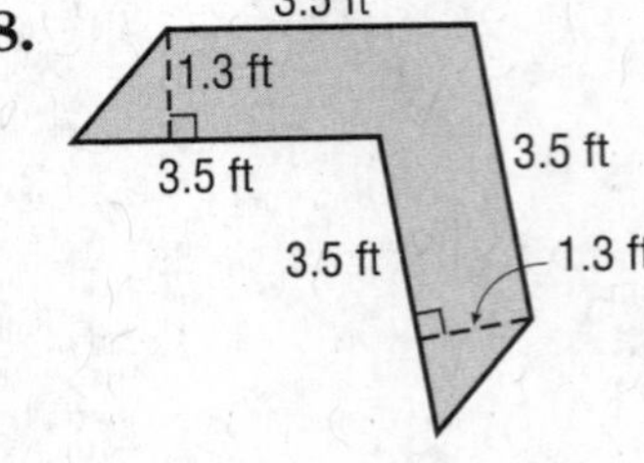

NAME ______________________ DATE __________ PERIOD ______

Reteach

Problem-Solving Investigation: Make a Model

When solving problems, one strategy that is helpful is to make a model. If a problem gives data that can be displayed visually, it may be useful to make a model of the situation. The model can then be used in order to solve the problem.

You can use the *make a model* strategy, along with the following four-step problem solving plan to solve a problem.

1 Understand – Read and get a general understanding of the problem.

2 Plan – Make a plan to solve the problem and estimate the solution.

3 Solve – Use your plan to solve the problem.

4 Check – Check the reasonableness of your solution.

Example DISPLAYS **A grocery store employee is making a pyramid display of boxes of a new cereal. If he doesn't want to have more than 4 rows in his display, what is the least number of cereal boxes he can use?**

Understand The cereal boxes need to be stacked in the shape of a pyramid. There should only be 4 rows in the pyramid. We need to know the minimum number of boxes of cereal needed to make a pyramid.

Plan Make a model to find the number of cereal boxes needed.

Solve Use a rectangle to represent each cereal box.

The least number of boxes needed is 4 + 3 + 2 + 1, or 10 boxes.

Check Count the number of boxes in the model. There are 10 boxes.

Exercise

TILING Demarcus has 18 decorative square tiles to make a design on a kitchen backsplash. He wants to arrange them in a rectangular shape with the least perimeter possible. How many tiles will be in each row?

NAME ______________________ DATE ____________ PERIOD _____

Skills Practice

Problem-Solving Investigation: Make a Model

Use the *make a model* strategy to solve.

1. **PATIO** Jarnail has 24 square brick pavers to arrange for a small patio to place his grill. He wants to place them in a rectangular shape with the least perimeter possible. How many bricks will be in each row?

2. **CRAFTS** Nyah is making a collage of her friends' school pictures on a poster board. Each picture is 2 inches by 3 inches and the poster board is 8 inches by 16 inches. What are the most pictures that Nyah can fit on the poster board if none of them overlap and all the pictures are facing the same direction?

3. **BOOKS** A bookstore arranges its best-seller books in the front window. In how many different ways can four best-seller books be arranged in a row?

4. **BASEBALLS** A sports store owner is making a display with 200 baseballs. He is placing them in layer as if to form a square pyramid. The bottom layer has 64 baseballs placed in the shape of a square. For each consecutive layer of baseballs, one baseball is placed where 4 baseballs meet. How many layers will be in the pyramid? How many baseballs will be left over?

NAME ______________________ DATE ____________ PERIOD ______

Reteach

Volume of Rectangular Prisms

The amount of space inside a three-dimensional figure is the **volume** of the figure. Volume is measured in **cubic units**. This tells you the number of cubes of a given size it will take to fill the prism.

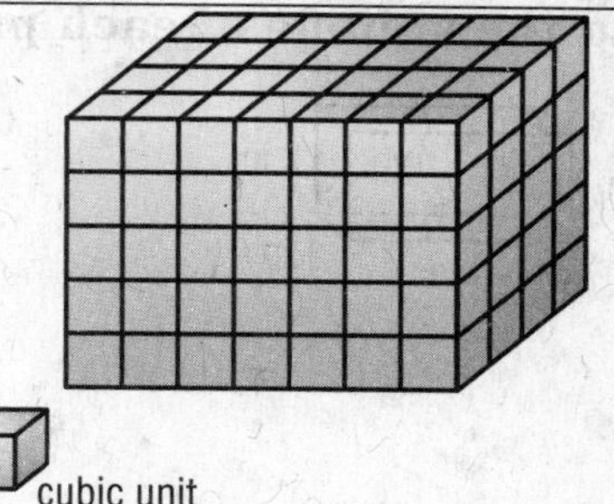

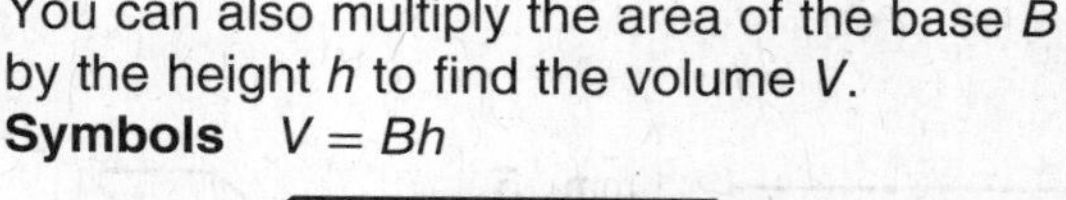

The volume V of a rectangular prism is the product of its length ℓ, width w, and height h.
Symbols $V = \ell wh$

Model

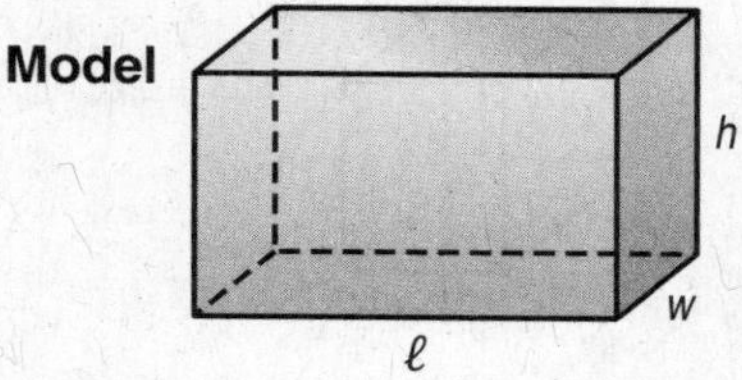

You can also multiply the area of the base B by the height h to find the volume V.
Symbols $V = Bh$

Model

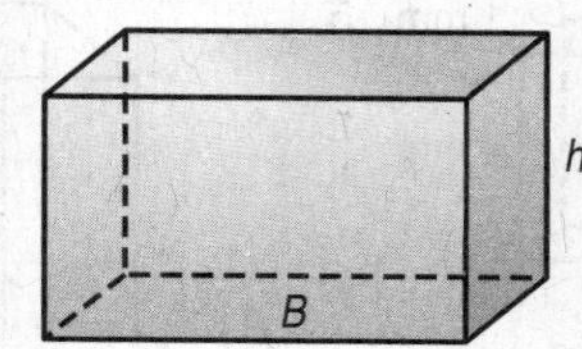

Example **Find the volume of the rectangular prism.**

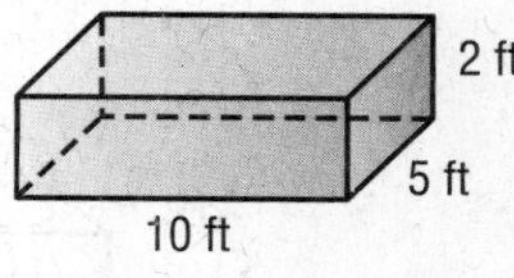

Method 1 Use $V = \ell wh$.

$V = \ell wh$

$V = 10 \times 5 \times 2$

$V = 100$

The volume is 100 ft^3.

Method 2 Use $V = Bh$.

$V = Bh$

$V = 50 \times 2$

$V = 100$

The volume is 100 ft^3.

Exercises

Find the volume of each prism.

1.

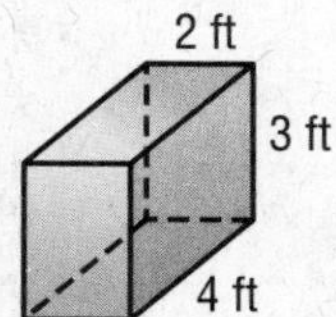

2.

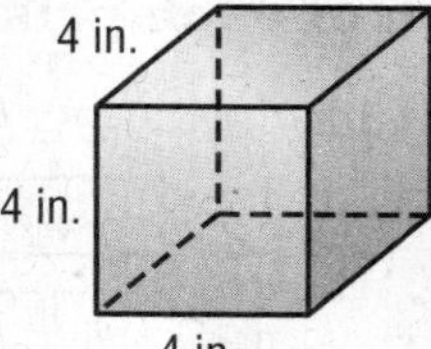

3.

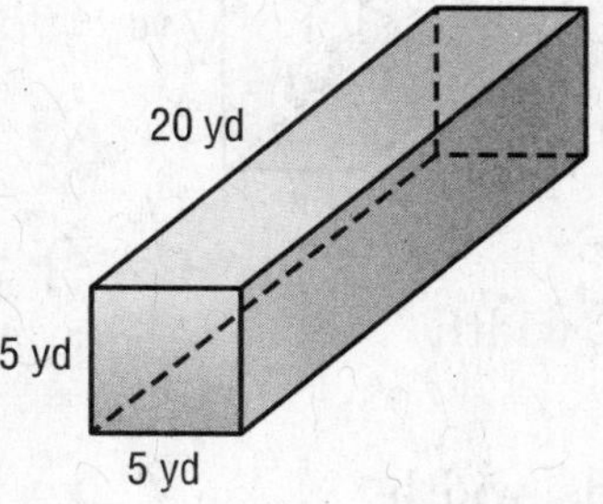

4.

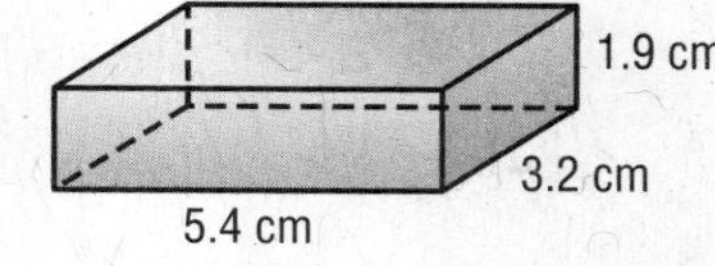

NAME ______________________ DATE ____________ PERIOD _____

Skills Practice

Volume of Rectangular Prisms

Find the volume of each prism.

1.

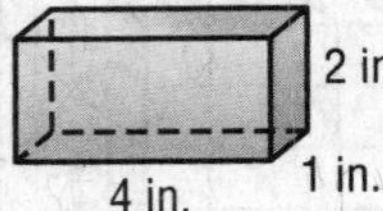

2.

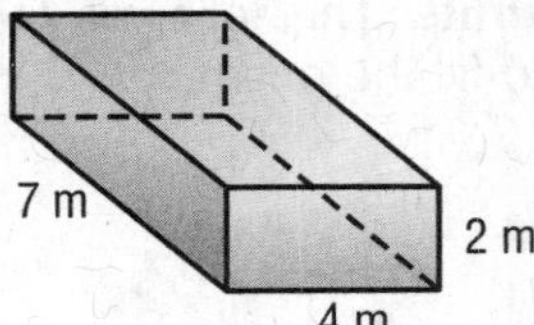

3.

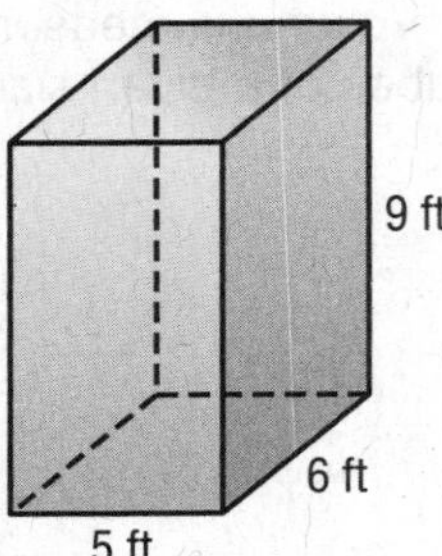

4.

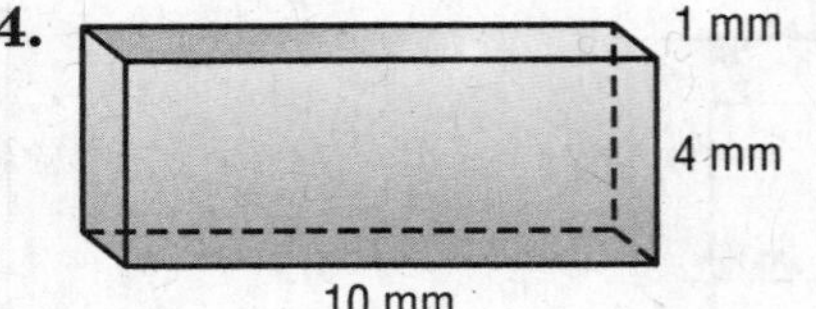

5.

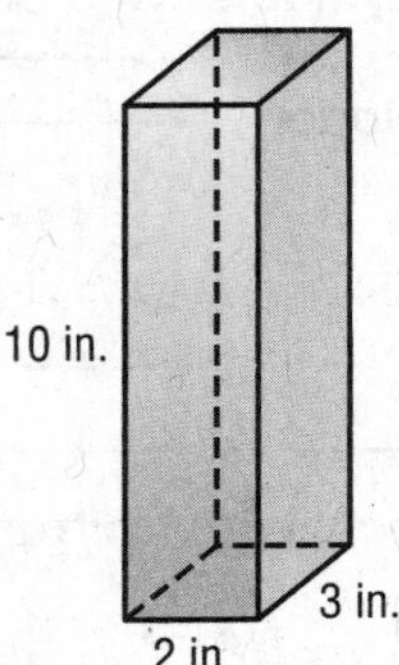

6.

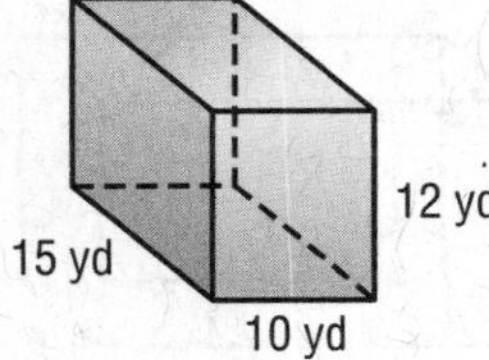

7.

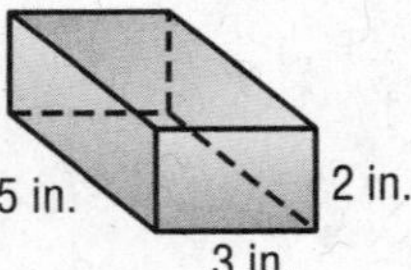

8.

9.

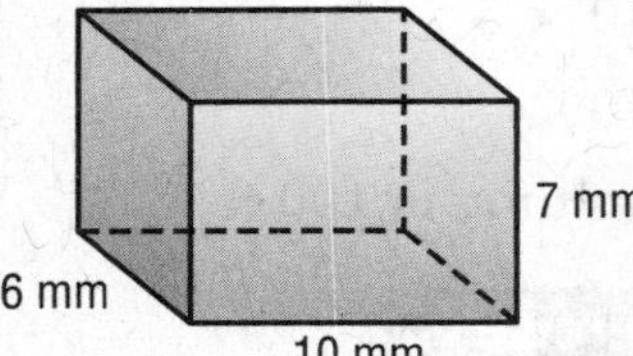

Find the missing dimension of each prism.

10. $V = 39.48\ m^3$

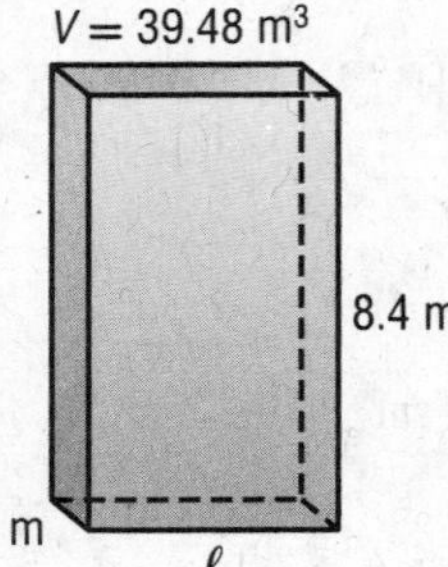

11. $V = 56\frac{1}{4}\ ft^3$

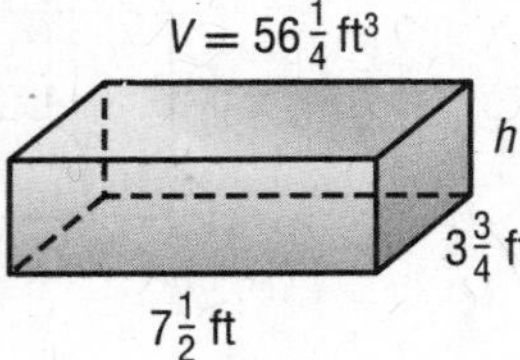

12. $V = 189\ yd^3$

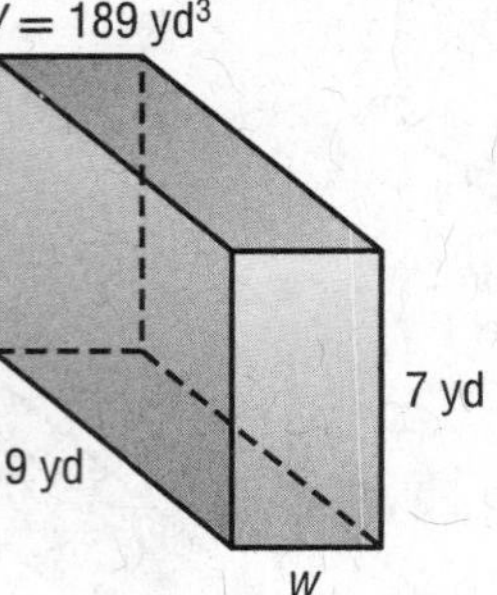

13. Find the volume of a rectangular prism with length 9 meters, width 4 meters, and height 5 meters.

14. What is the volume of a rectangular prism with length 6 yards, width 3 yards, and a height of 2 yards?

NAME ________________ DATE ________ PERIOD ______

Reteach

Surface Area of Rectangular Prisms

The **surface area** *S.A.* of a rectangular prism with length ℓ, width w, and height h is the sum of the areas of the faces.

Symbols $S.A. = 2\ell h + 2\ell w + 2hw$

Model

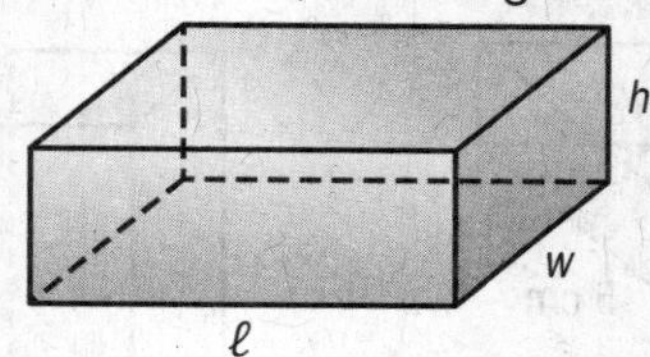

Example **Find the surface area of the rectangular prism.**

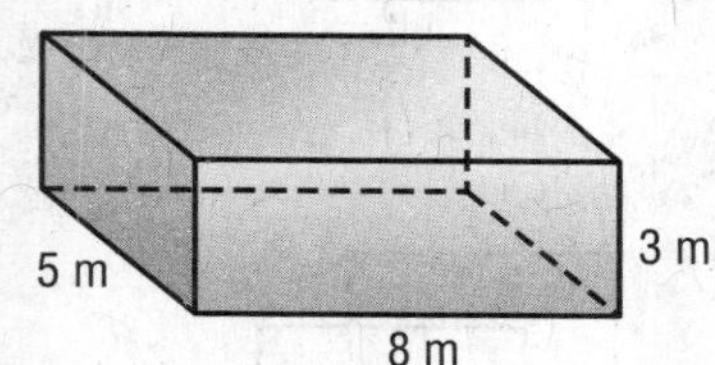

Find the area of each face.

front and back
$2\ell h = 2(8)(3) = 48$

top and bottom
$2\ell w = 2(8)(5) = 80$

two sides
$2hw = 2(3)(5) = 30$

Add to find the surface area. The surface area is 48 + 80 + 30 or 158 square meters.

Exercises

Find the surface area of each rectangular prism.

1.
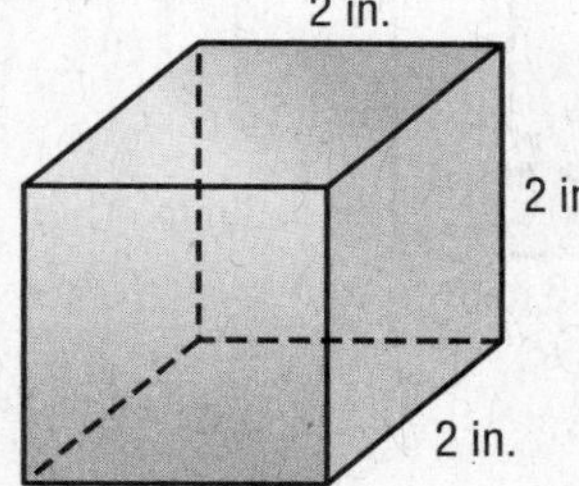

2.
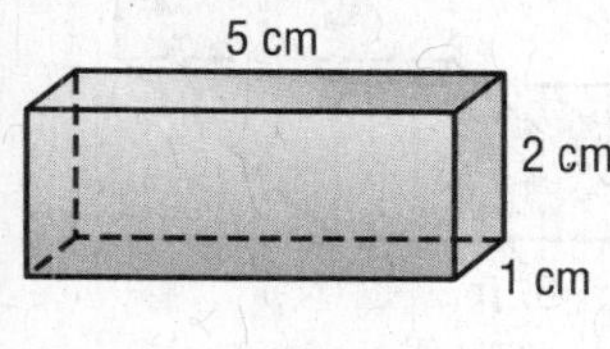

3.
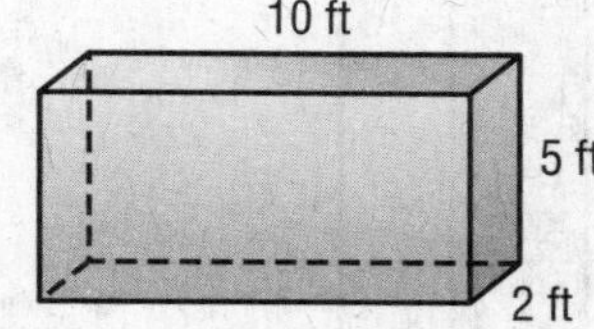

4.
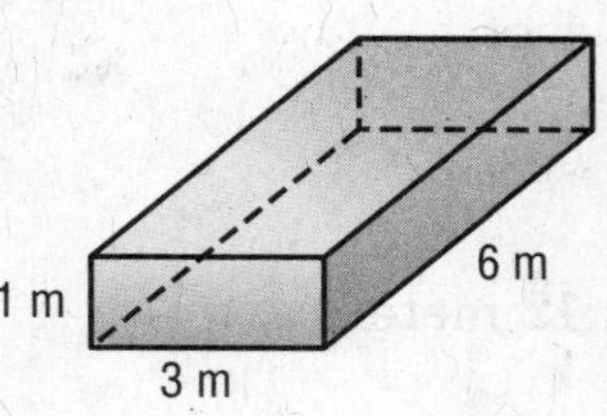

5.
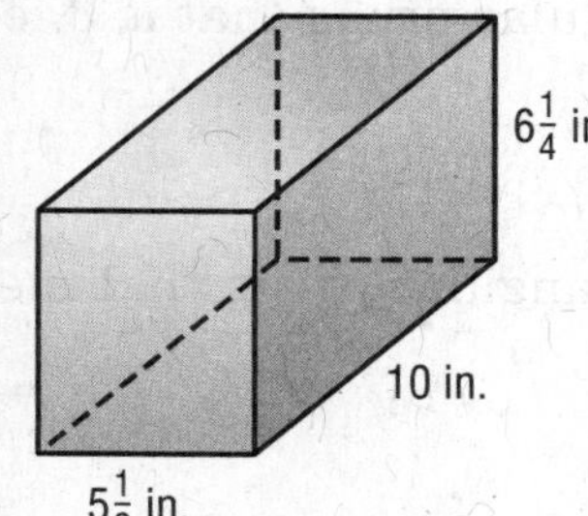

6.
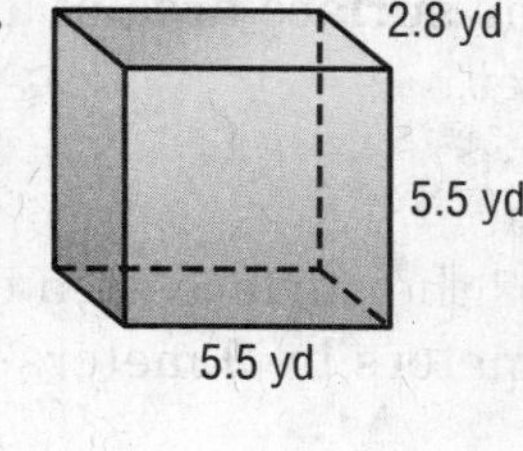

NAME ______________________ DATE ____________ PERIOD ______

Skills Practice

Surface Area of Rectangular Prisms

Find the surface area of each rectangular prism.

1.

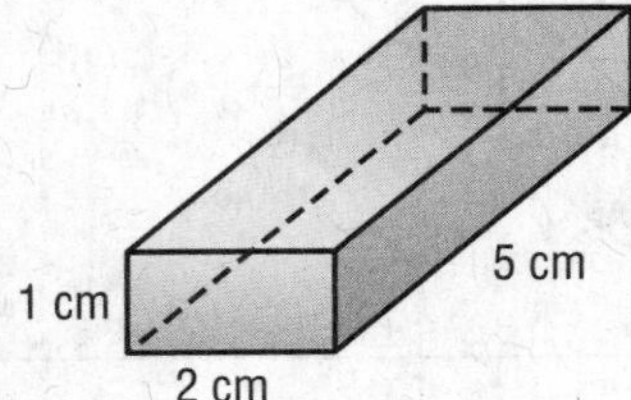

2.

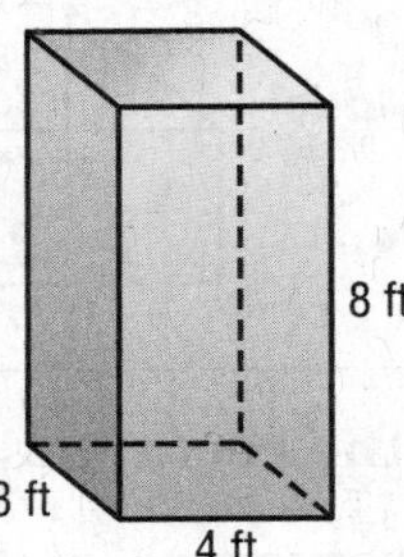

3.

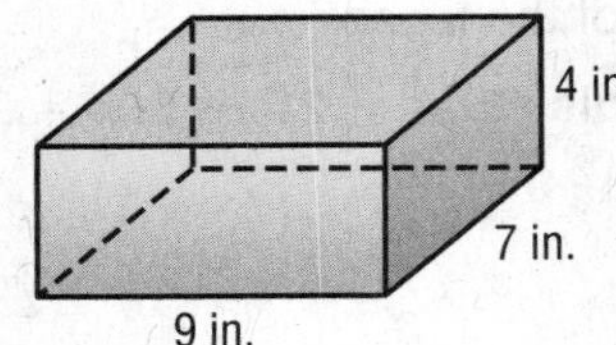

4.

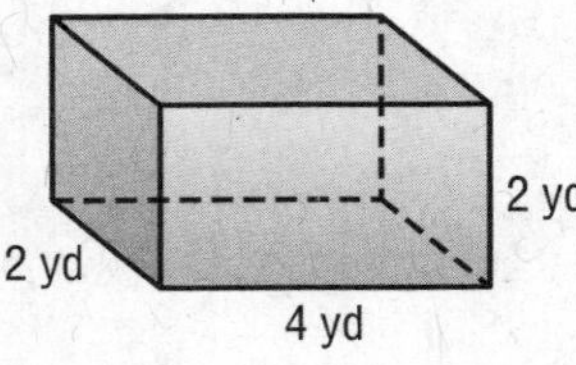

5.

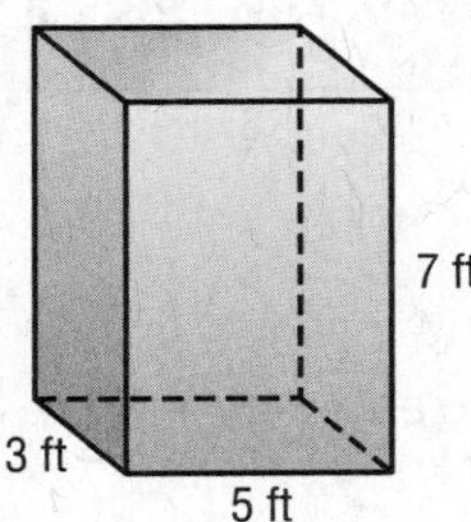

6.

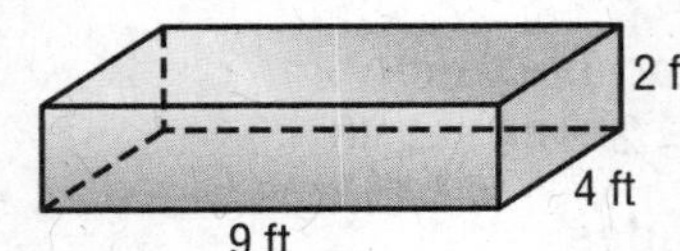

7.

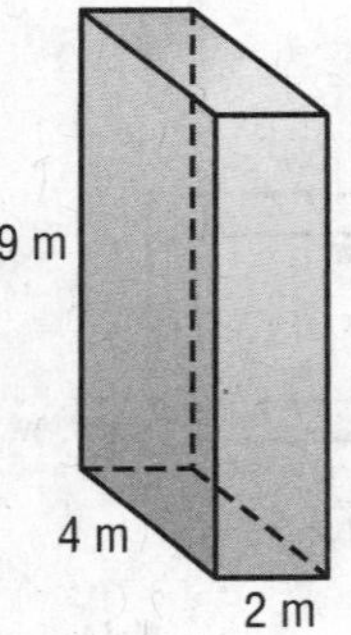

8.

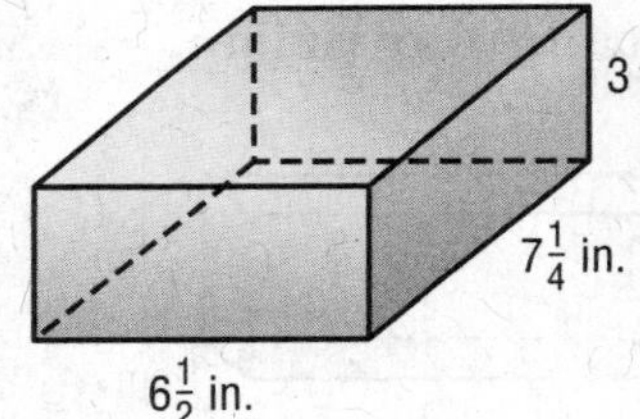

9.

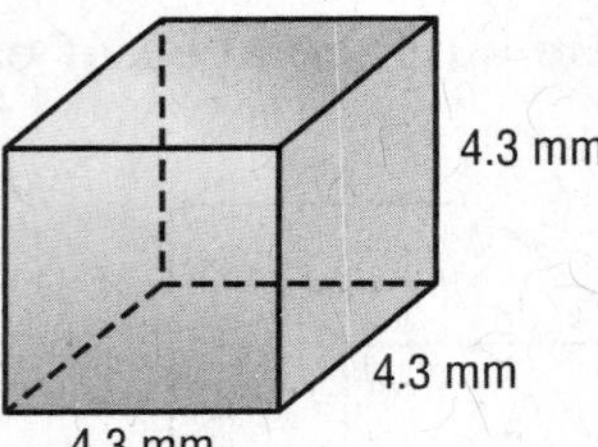

10. Find the surface area of a rectangular prism that is 3 feet by 4 feet by 6 feet.

11. What is the surface area of a rectangular prism that measures 12 meters by 11 meters by 9 meters?

NAME ______________________ DATE ____________ PERIOD ______

Reteach

Volume of Triangular Prisms

Volume of a Triangular Prism	
Words The volume V of a triangular prism is the area of the base B times the height h.	**Model** 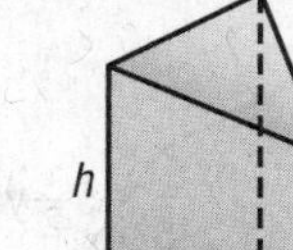
Symbols $V = Bh$, where $B = \frac{1}{2}bh$	

Example 1 **Find the volume of the triangular prism.**

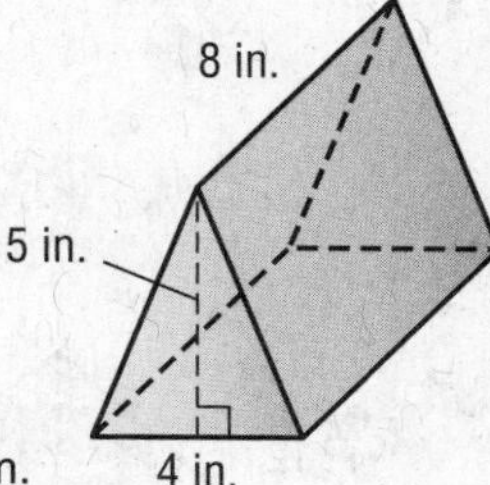

The area of the triangle is $\frac{1}{2} \cdot 4 \cdot 5$, so replace B with $\frac{1}{2} \cdot 4 \cdot 5$.

$V = Bh$ — Volume of a prism

$V = \left(\frac{1}{2} \cdot 4 \cdot 5\right)(h)$ — Replace B with $\frac{1}{2} \cdot 4 \cdot 5$.

$V = \left(\frac{1}{2} \cdot 4 \cdot 5\right)(8)$ — Replace h with 8, the height of the prism.

$V = 80$ — Multiply.

The volume is 80 cubic inches or 80 in^3.

Example 2 **Find the volume of the triangular prism.**

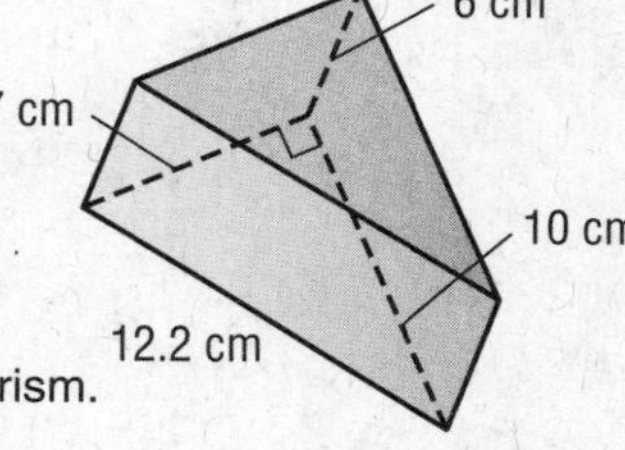

$V = Bh$ — Volume of a prism

$V = \left(\frac{1}{2} \cdot 7 \cdot 10\right)(h)$ — Replace B with $\frac{1}{2} \cdot 7 \cdot 10$.

$V = \left(\frac{1}{2} \cdot 7 \cdot 10\right)(6)$ — Replace h with 6, the height of the prism.

$V = 210$ — Multiply.

The volume is 210 cubic centimeters or 210 cm^3.

Exercises

Find the volume of each prism. Round to the nearest tenth if necessary.

1.

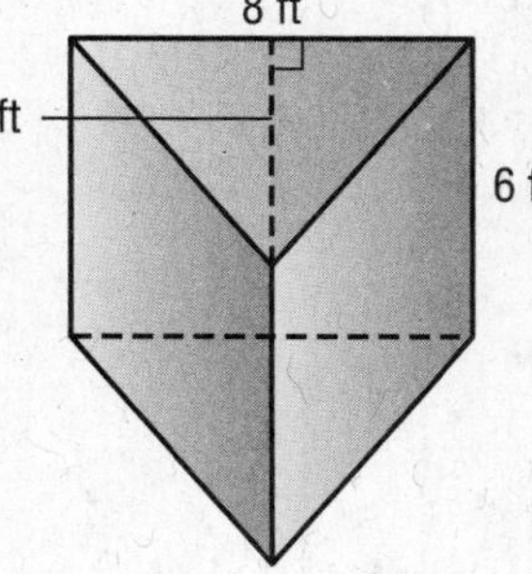

2.

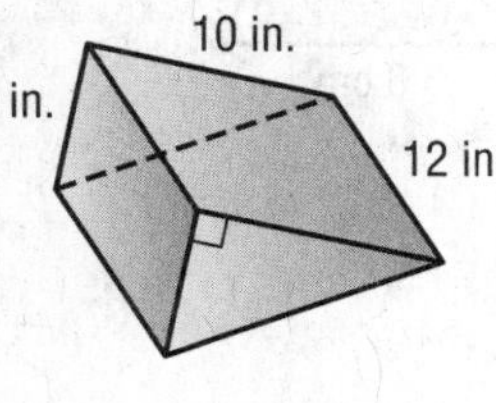

3.

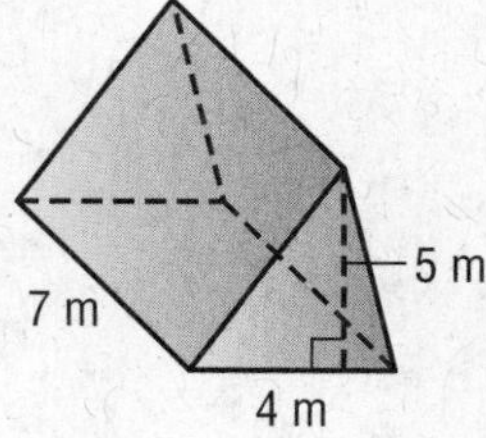

NAME ______________________ DATE ____________ PERIOD ______

Skills Practice

Volume of Triangular Prisms

Find the volume of each prism. Round to the nearest tenth if necessary.

1.

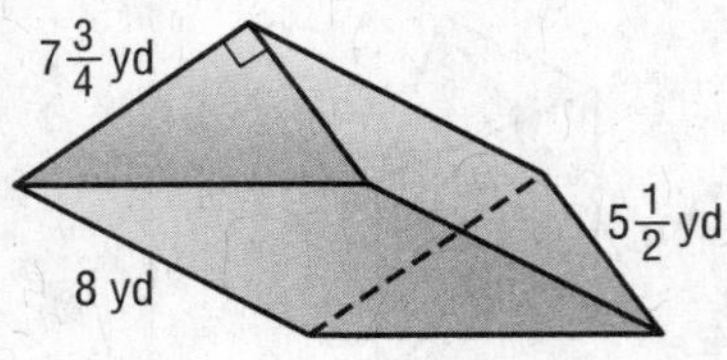

2.

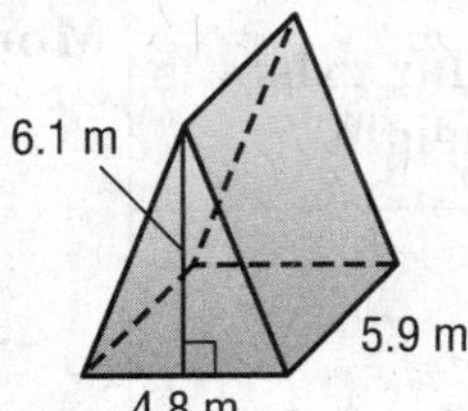

3.

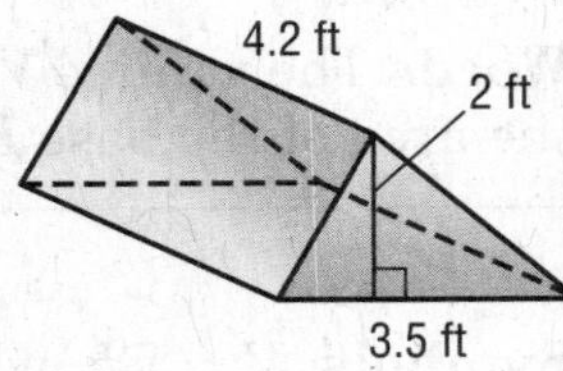

4.

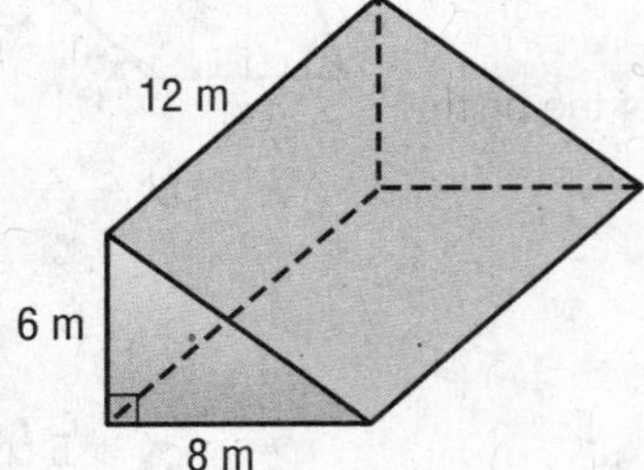

5.

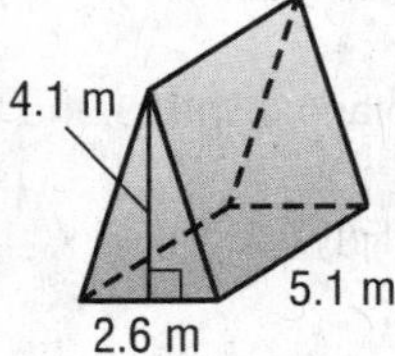

6.

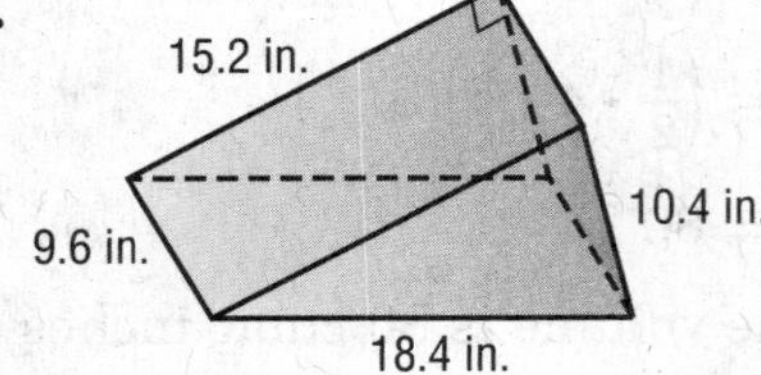

7.

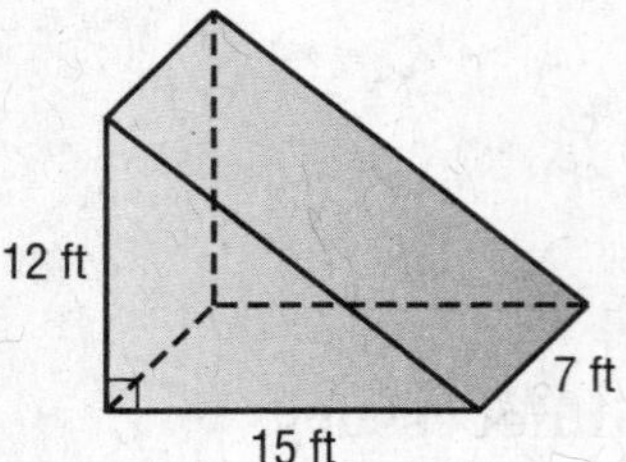

8.

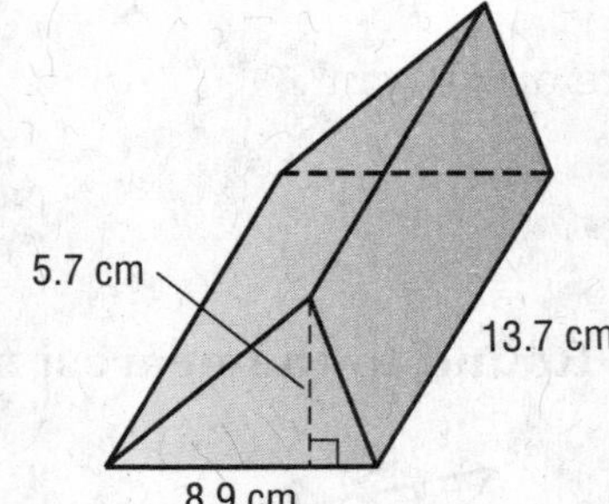

9.

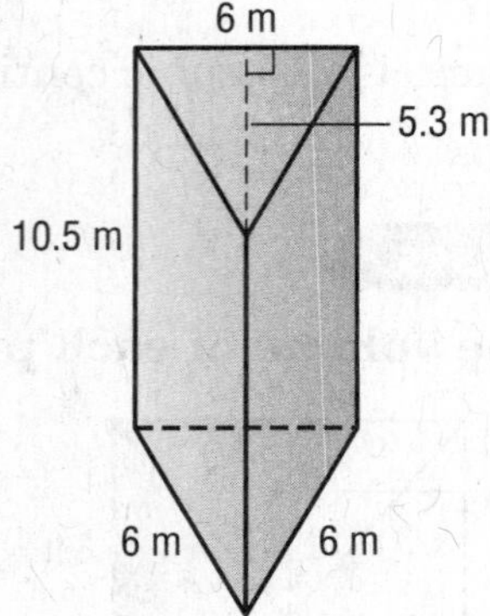

NAME ______________________ DATE ____________ PERIOD ______

Reteach

Volume of Pyramids

Volume of a Pyramid	
Words The volume V of a pyramid is one third the area of the base B times the height h.	**Model** 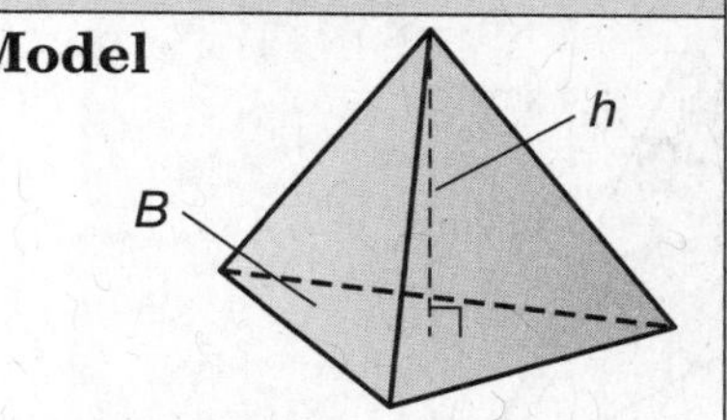
Symbols $V = \frac{1}{3}Bh$	

Example 1 **Find the volume of the pyramid.**

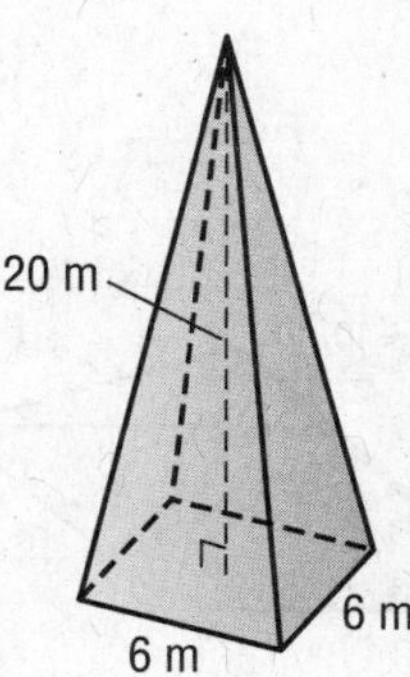

$V = \frac{1}{3}Bh$ Volume of a pyramid

$V = \frac{1}{3}(6 \cdot 6)(20)$ The base is a square, so replace B with $6 \cdot 6$ and h with 20.

$V = 240$ Multiply.

The volume is 240 cubic meters or 240 m^3.

Example 2 **Find the volume of the pyramid.**

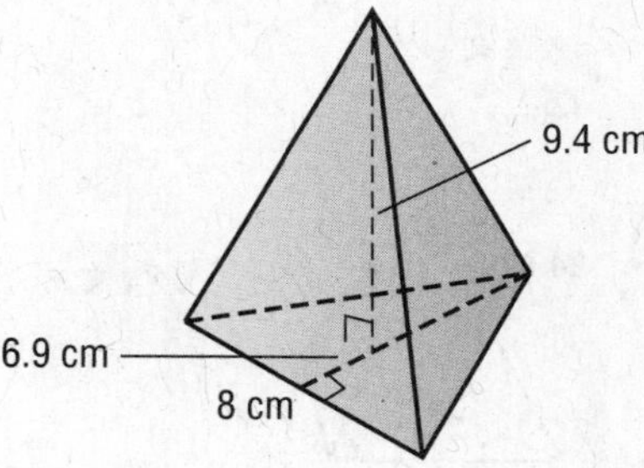

$V = \frac{1}{3}Bh$ Volume of a pyramid

$V = \frac{1}{3}(27.6)(9.4)$ The base is a triangle, so replace B with $\frac{1}{2} \cdot 8 \cdot 6.9$ or 27.6, and h with 9.4.

$V = 86.48$ Multiply.

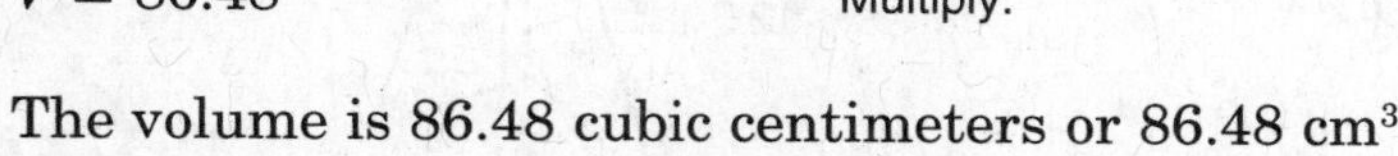

The volume is 86.48 cubic centimeters or 86.48 cm^3.

Exercises

Find the volume of each pyramid. Round to the nearest tenth if necessary.

1. 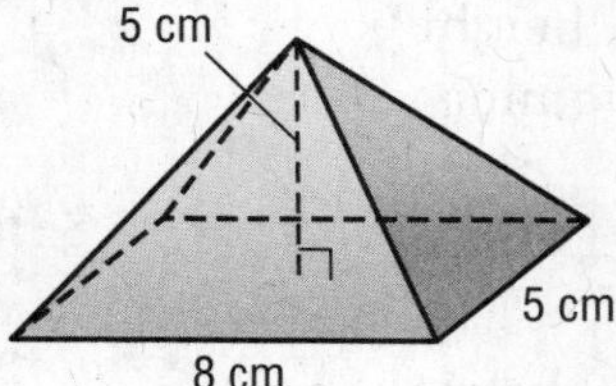

2. 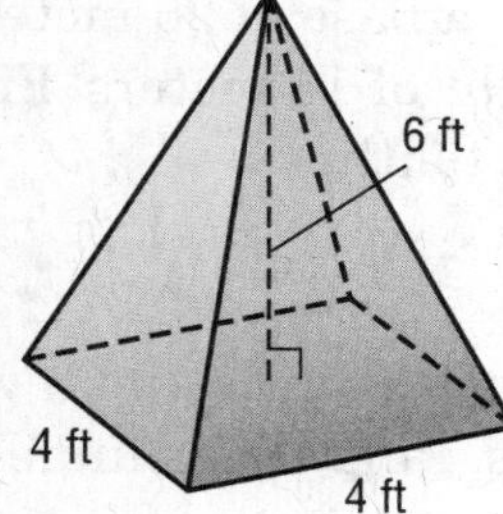

3. 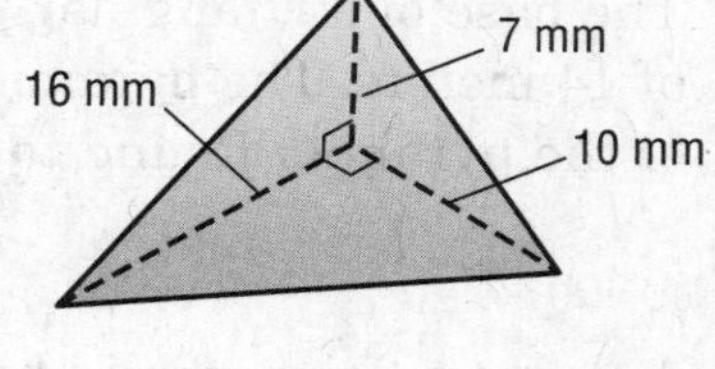

NAME ______________________ DATE ____________ PERIOD ______

Skills Practice

Volume of Pyramids

Find the volume of each pyramid. Round to the nearest tenth if necessary.

1.

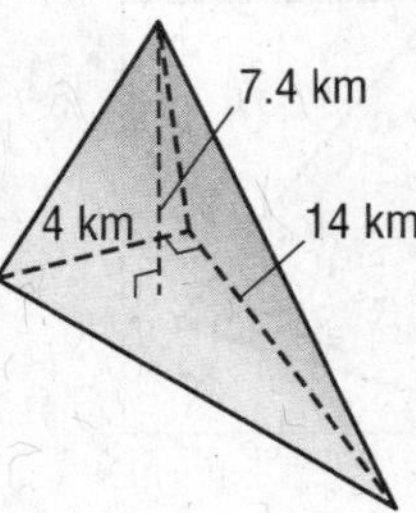

2.

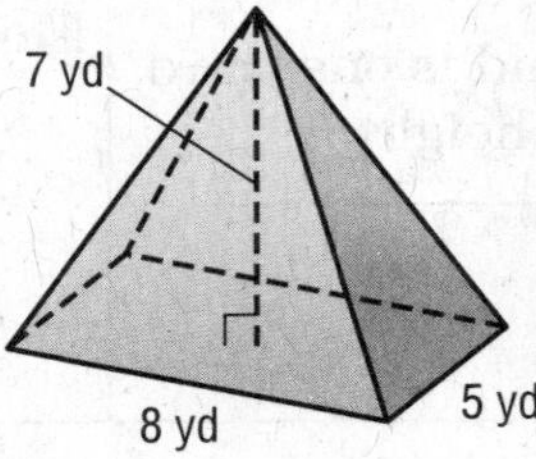

3.

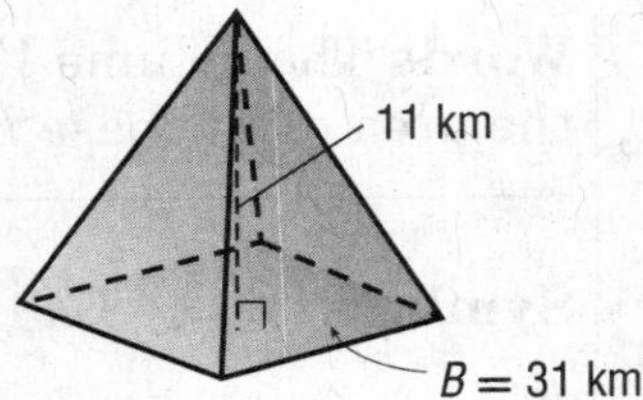

4.

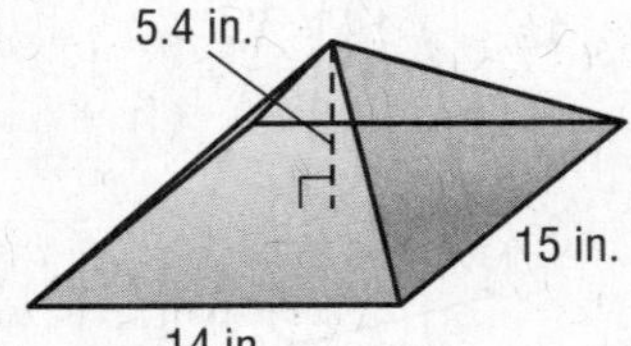

5.

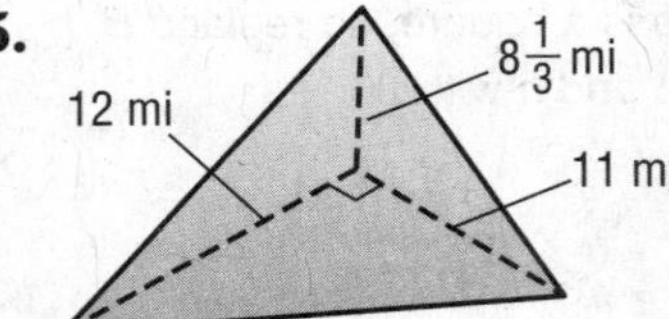

6.

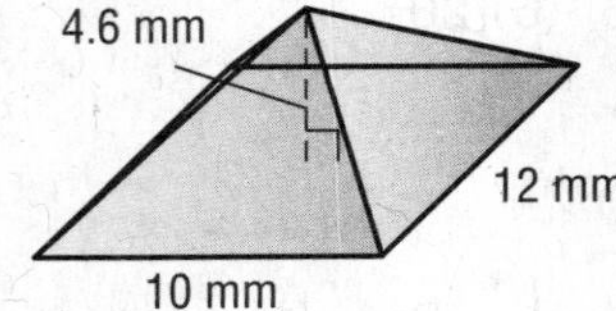

7.

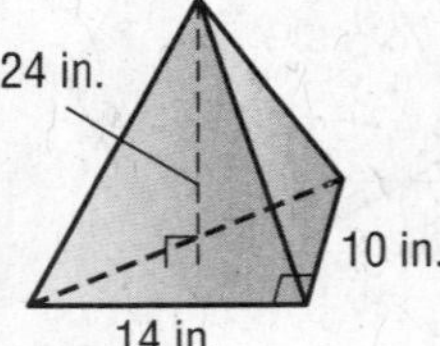

8.

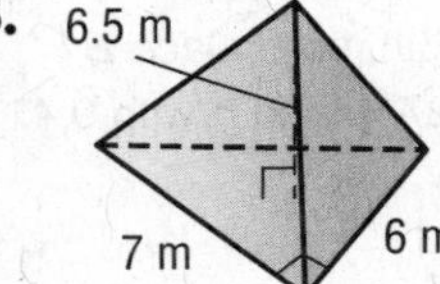

9.

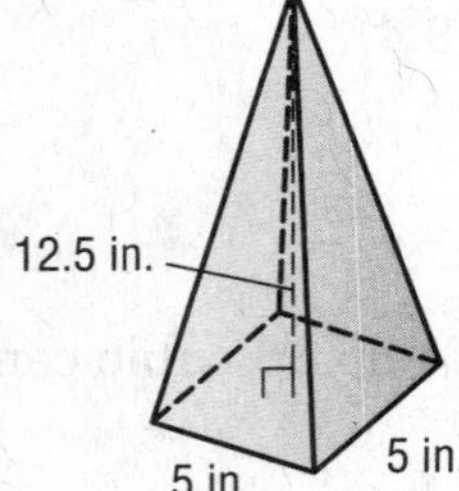

10. The base of a triangular pyramid has a base of 20 meters and a height of 14 meters. The pyramid has a height of 14 meters. Find the volume of the pyramid. Round to the nearest tenth.

11. A rectangular pyramid has a length of 19 yards and a width of 8 yards. The pyramid has a height of 15 yards. Find the volume of the pyramid.

NAME ______________________ DATE ____________ PERIOD _____

Reteach

Volume of Cylinders

As with prisms, the area of the base of a **cylinder** tells the number of cubic units in one layer. The height tells how many layers there are in the cylinder. The volume V of a cylinder with radius r is the area of the base B times the height h.

$V = Bh$ or $V = \pi r^2 h$, where $B = \pi r^2$

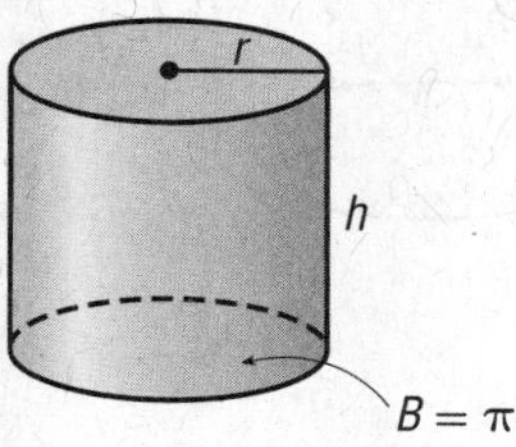

Example **Find the volume of the cylinder. Use 3.14 for π. Round to the nearest tenth.**

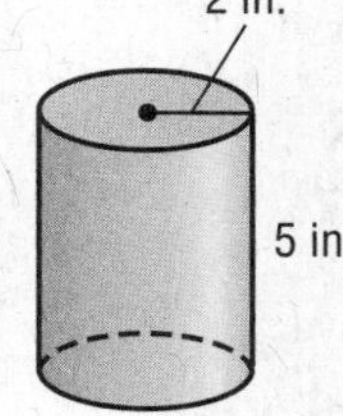

$V = \pi r^2 h$	Volume of a cylinder
$V \approx 3.14(2)^2(5)$	Replace π with 3.14, r with 2, and h with 5.
$V \approx 62.8$	Multiply.

The volume is about 62.8 cubic inches.

Exercises

Find the volume of each cylinder. Use 3.14 for π. Round to the nearest tenth.

1.

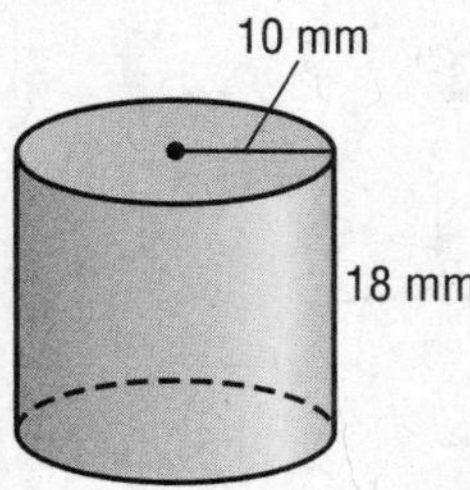

2.

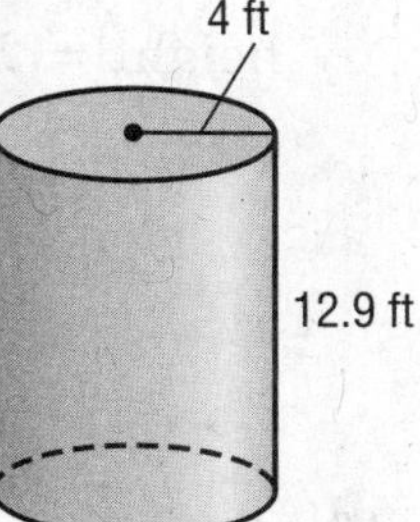

3.

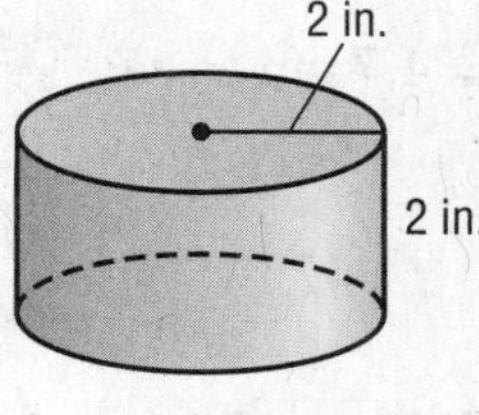

4. radius = 9.5 yd
height = 2.2 yd

5. diameter = 6 cm
height = 11 cm

6. diameter = 3 m
height = 1 m

NAME ______________________ DATE ____________ PERIOD ______

Skills Practice

Volume of Cylinders

Find the volume of each cylinder. Use 3.14 for π. Round to the nearest tenth.

1.

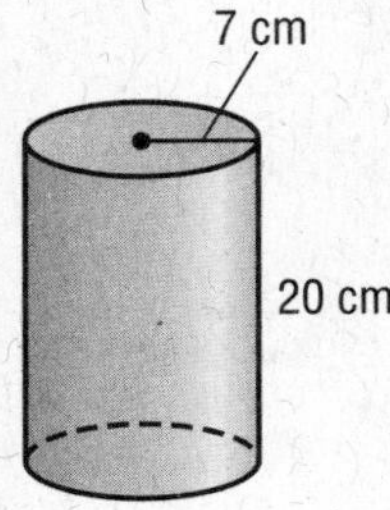

2.

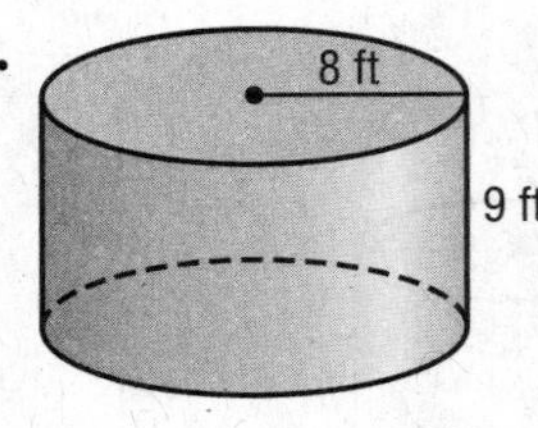

3.

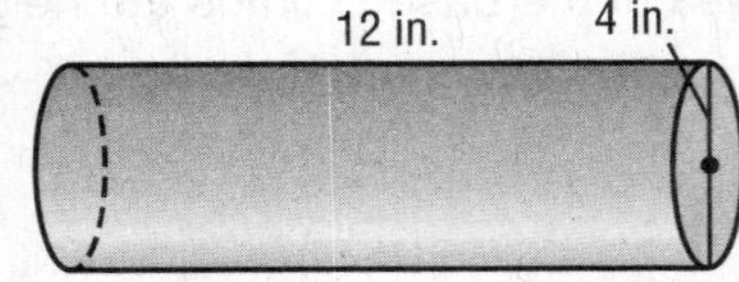

4.

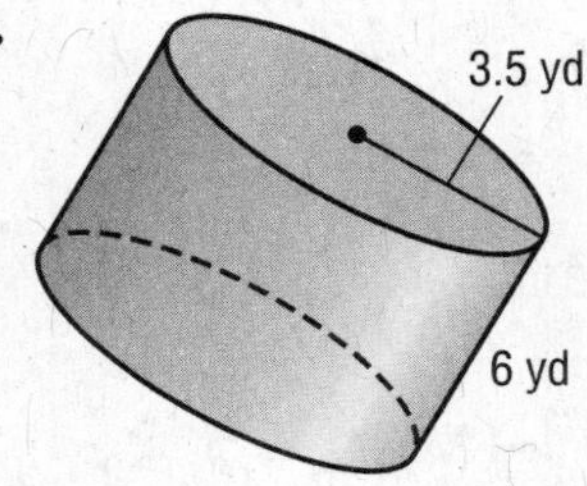

5.

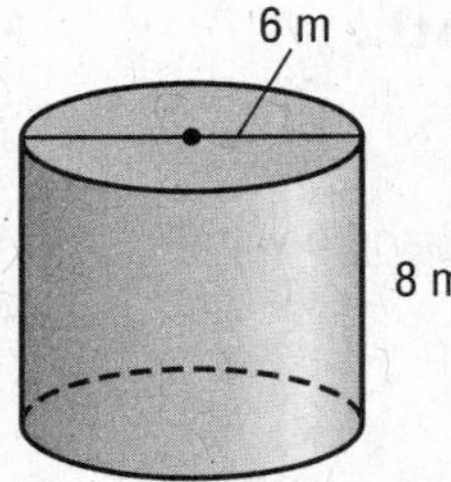

6.

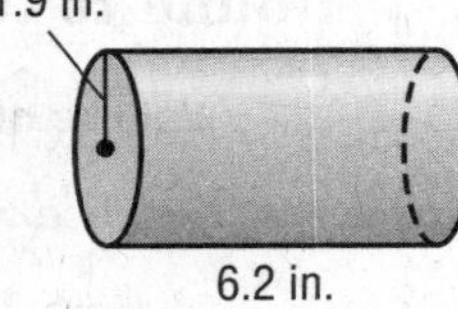

7. radius = 10 cm
height = 4.7 cm

8. radius = 4 ft
height = $2\frac{1}{2}$ ft

9. diameter = 10 mm
height = 4 mm

10. diameter = 7.1 in.
height = 1 in.

Multi-Part Lesson 2 PART D

NAME ______________________ DATE ____________ PERIOD ______

Reteach

Volume of Cones

Volume of a Cone	
Words The volume V of a cone with radius r is one third the area of the base B times the height h.	**Model** 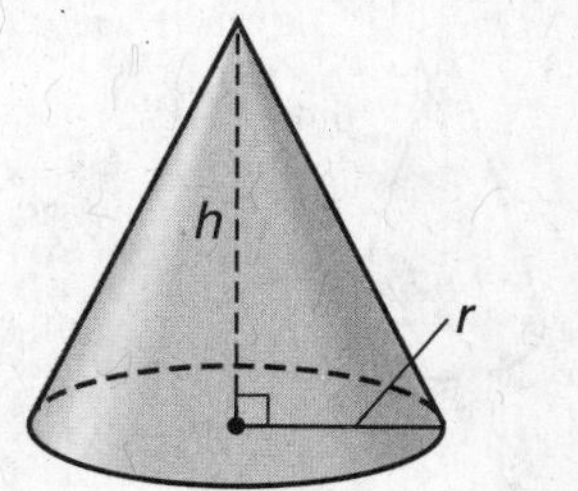
Symbols $V = \frac{1}{3}Bh$ or $V = \frac{1}{3}\pi r^2 h$	

Example 1 **Find the volume of the cone. Use 3.14 for π. Round to the nearest tenth.**

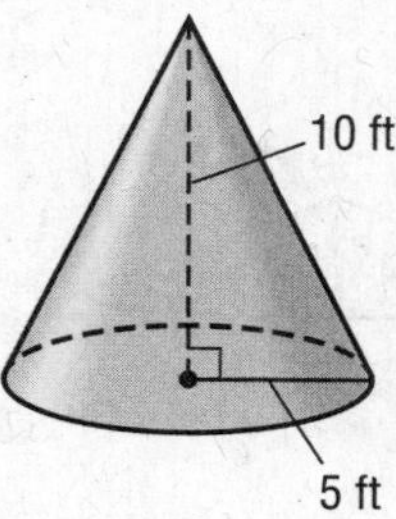

$V = \frac{1}{3}\pi r^2 h$ Volume of a cone

$V \approx \frac{1}{3} \cdot 3.14 \cdot 5 \cdot 5 \cdot 10$ $\pi \approx 3.14$, $r = 5$, $h = 10$

$V \approx 261.7$ Simplify.

The volume is about 261.7 cubic feet.

Example 2 **Find the volume of the cone. Use 3.14 for π. Round to the nearest tenth.**

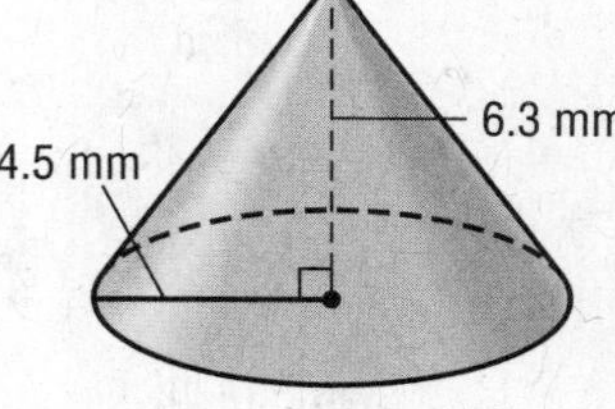

$V = \frac{1}{3}\pi r^2 h$ Volume of a cone

$V \approx \frac{1}{3} \cdot 3.14 \cdot 4.5 \cdot 4.5 \cdot 6.3$ $\pi \approx 3.14$, $r = 4.5$, $h = 6.3$

$V \approx 133.5$ Simplify.

The volume is about 133.5 cubic millimeters.

Exercises

Find the volume of each cone. Use 3.14 for π. Round to the nearest tenth if necessary.

1.

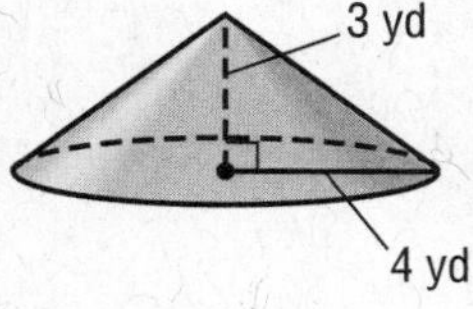

2.

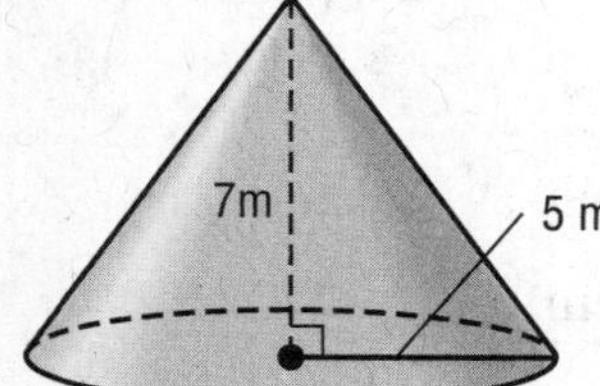

3.

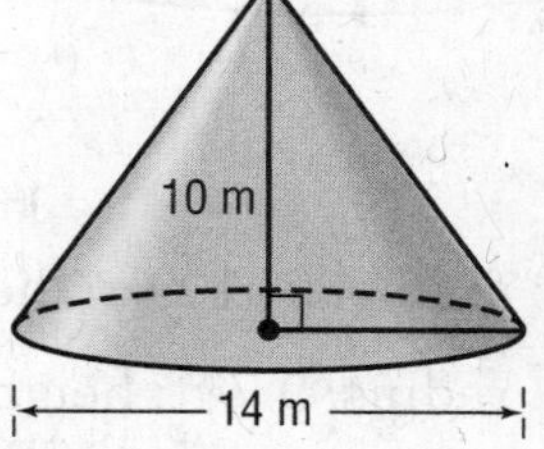

NAME ______________________________ DATE ____________ PERIOD _____

Skills Practice

Volume of Cones

Find the volume of each cone. Use 3.14 for π. Round to the nearest tenth.

1.

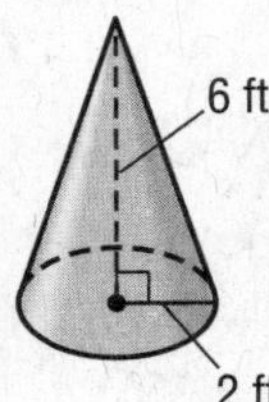

2.

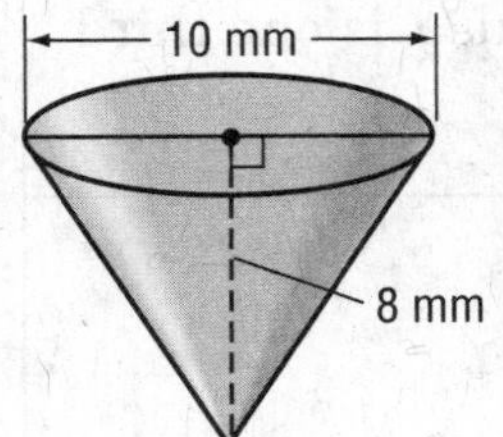

3.

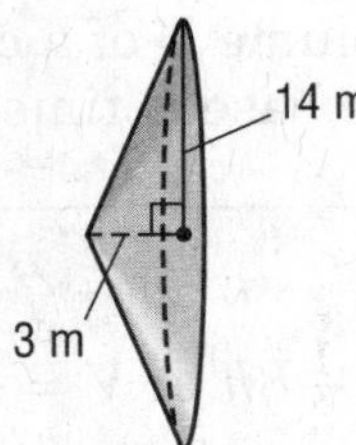

4.

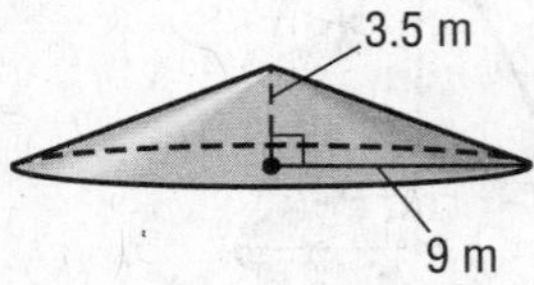

5.

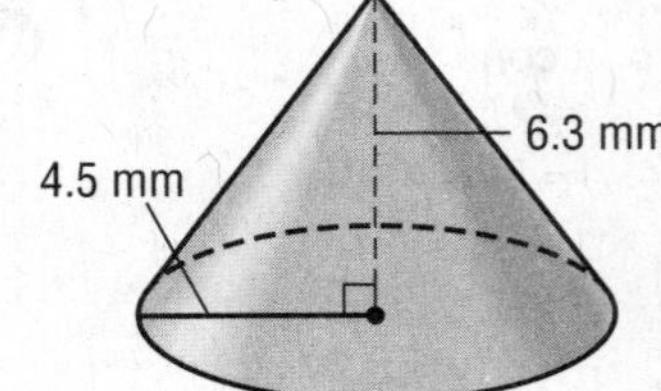

6.

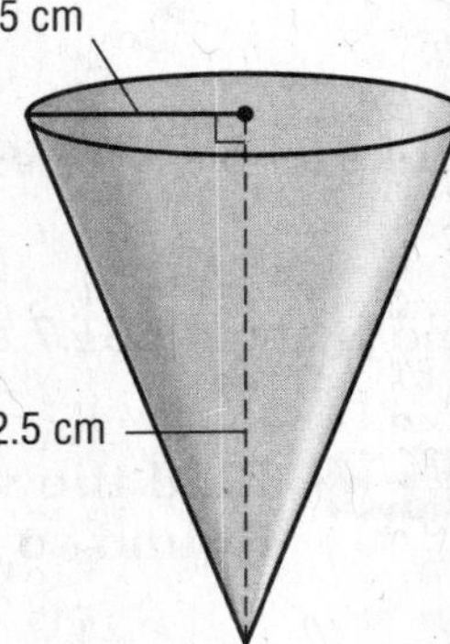

7.

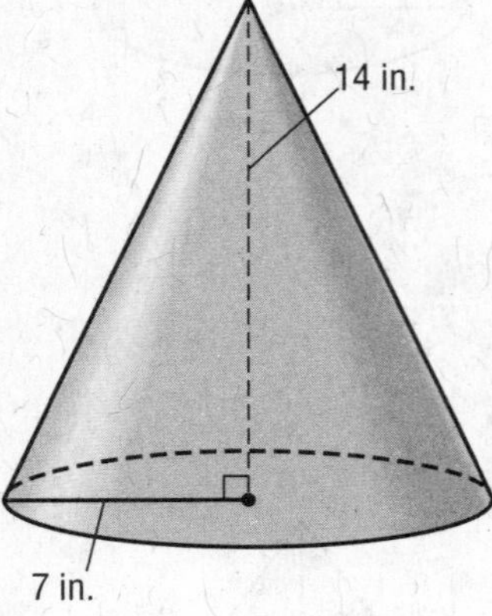

8.

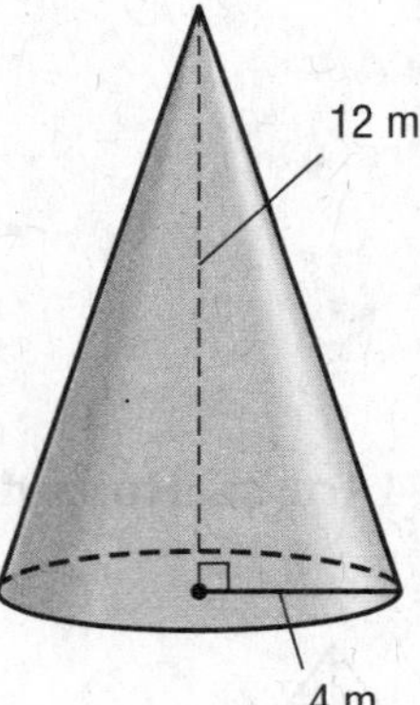

9.

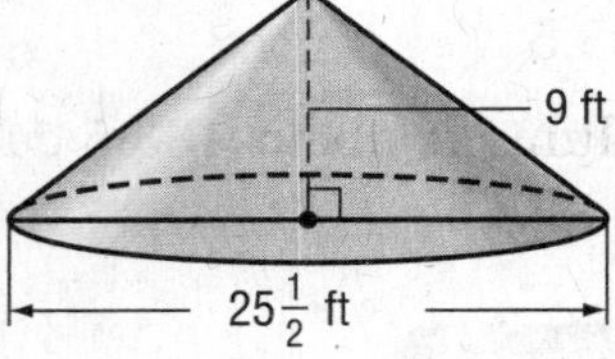

10. cone: diameter, 10 cm; height, 12 cm

11. cone: radius, 9.7 ft; height, 18 ft

12. A cone with a radius of 16 centimeters has a volume of 3,215.36 cubic centimeters. Find the height of the cone. Use 3.14 for π.

NAME _______________ DATE ________ PERIOD _____

Reteach

Problem-Solving Investigation: Draw a Diagram

When solving problems, one strategy that is helpful is to draw a diagram. A problem may often describe a situation that is easier to solve visually.

You can draw a diagram, along with the following four-step problem-solving plan to solve a problem.

1 Understand – Read and get a general understanding of the problem.

2 Plan – Make a plan to solve the problem and estimate the solution.

3 Solve – Use your plan to solve the problem.

4 Check – Check the reasonableness of your solution.

Example **Elise's mother wants her to cover an empty juice can with wallpaper so that she can use it to hold pens on her desk. The can has a height of 6 inches and a radius of 3 inches. Draw a diagram to determine how much wallpaper she will need to cover the can. The can will not have a lid. Use 3.14 for π.**

UNDERSTAND The can is a cylinder. You know its dimensions.

PLAN Make a diagram of the can without a lid.

SOLVE Find the area of the rectangle and the circle.

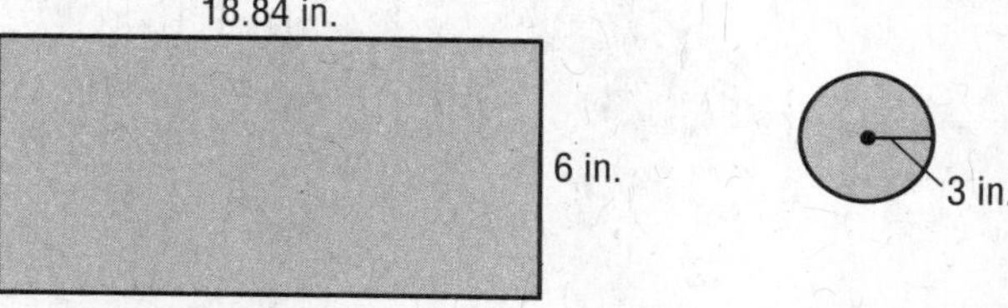

$6 \times 18.84 = 113.04$ $\pi(3)^2 \approx 28.3$

The surface area is about $113.04 + 28.3$ or 141.34 square inches.

CHECK Check for reasonableness. $20 \times 6 = 120$; $3(3)^2 = 27$; $120 + 27 = 147 \approx 141.34$

Exercise

BIRDHOUSE Tristan wants to cover the birdhouse with vinyl to protect it from the weather. The door has a diameter of 1.5 inches. Find the amount of vinyl Tristan will need to cover the birdhouse. Use 3.14 for π. Round to the nearest tenth.

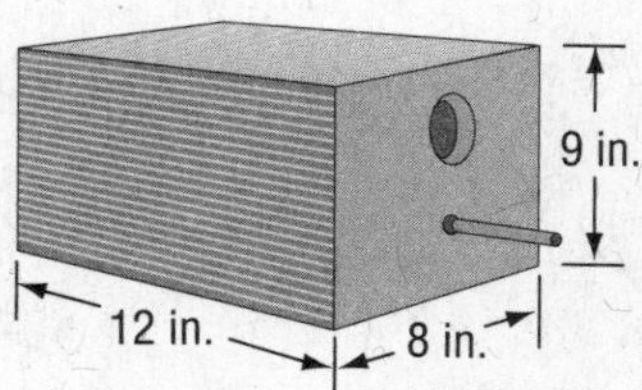

NAME ______________________ DATE ____________ PERIOD _____

Skills Practice

Problem-Solving Investigation: Draw a Diagram

Solve by using the *draw a diagram* strategy.

1. **DOG FOOD** A container used to keep dog food fresh is 30 inches tall by 18 inches long by 14 inches wide. Find the surface area of the dog food container, including the lid.

2. **DOORSTOP** A museum uses a brick from an old building as a doorstop. The brick is 3.5 inches tall by 4 inches wide by 8 inches long. Find the surface area of the brick.

3. **COLLECTION** A marble collector keeps his marbles in an empty paint can with a diameter of 7 inches and height of 8 inches. He wants to wrap the side of the can in brown paper so he can write facts about the marbles on it. Find the amount of brown paper he will need. Use 3.14 for π. Round to the nearest tenth.

4. **FOODS** A company that produces paper packaging sells a cylindrical container for salt. The container is 9 inches tall. The radius is one third the height. Find the amount of paper it will take to make the salt container. Use 3.14 for π. Round to the nearest tenth.

Multi-Part Lesson 3 PART C

NAME ______________________ DATE ____________ PERIOD ______

Reteach

Surface Area of Cylinders

Surface Area of a Cylinder	
Words The surface area *S.A.* of a cylinder with height h and radius r is the sum of the area of the curved surface and the areas of the circular bases.	**Model** 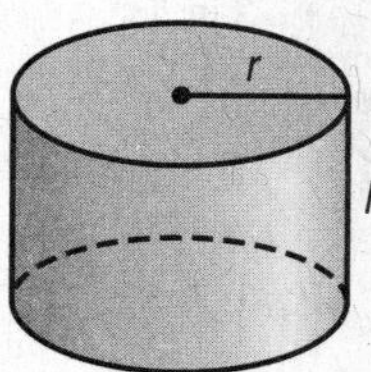
Symbols $S.A. = 2\pi rh + 2\pi r^2$	

Example 1 **Find the surface area of the cylinder. Use 3.14 for π. Round to the nearest tenth.**

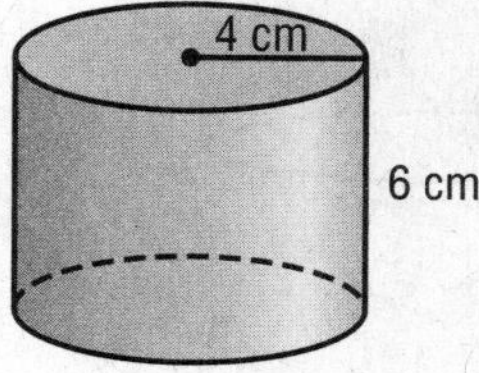

$S.A. = 2\pi rh + 2\pi r^2$	Surface area of a cylinder
$S.A. \approx 2(3.14)(4)(6) + 2(3.14)(4)^2$	Replace r with 4 and h with 6.
$S.A. \approx 251.2$	Multiply.

The surface area is about 251.2 square centimeters.

Example 2 **Find the surface area of the cylinder. Round to the nearest tenth if necessary.**

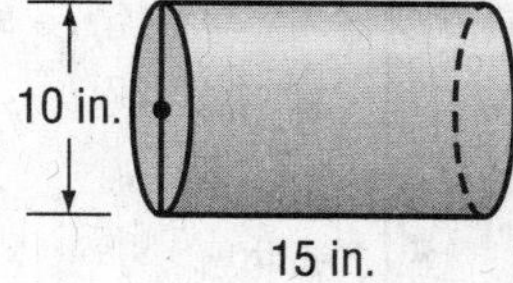

$S.A. = 2\pi rh + 2\pi r^2$	Surface area of a cylinder
$S.A. \approx 2(3.14)(5)(15) + 2(3.14)(5)^2$	Replace r with 5 and h with 15.
$S.A. \approx 628$	Multiply.

The surface area is about 628 square inches.

Exercises

Find the surface area of each cylinder. Use 3.14 for π. Round to the nearest tenth if necessary.

1.

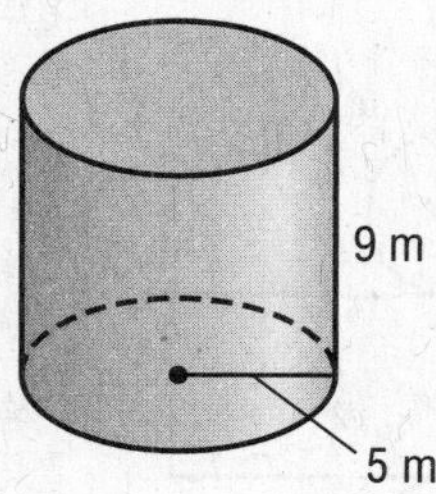

2.

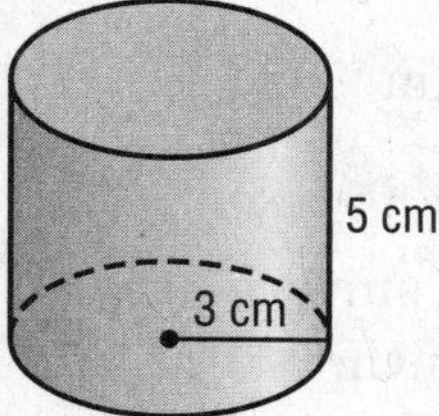

3.

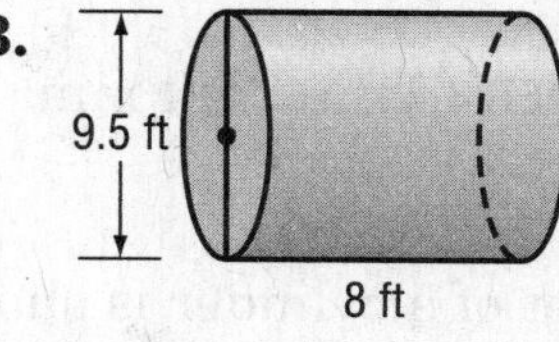

NAME ______________________ DATE ____________ PERIOD _____

Skills Practice

Surface Area of Cylinders

Find the surface area of each cylinder. Use 3.14 for π. Round to the nearest tenth if necessary.

1.

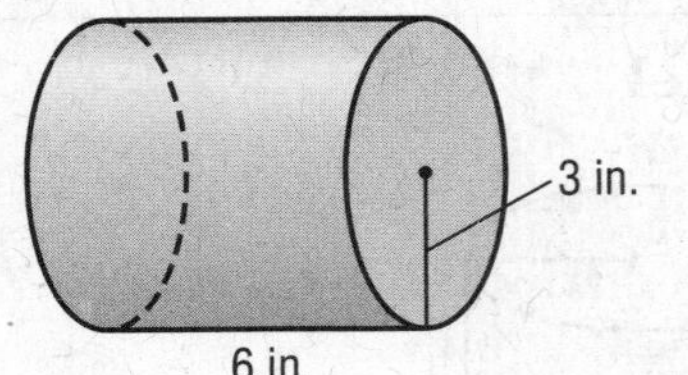

2.

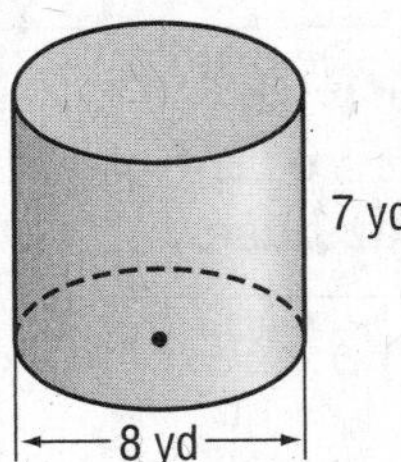

3.

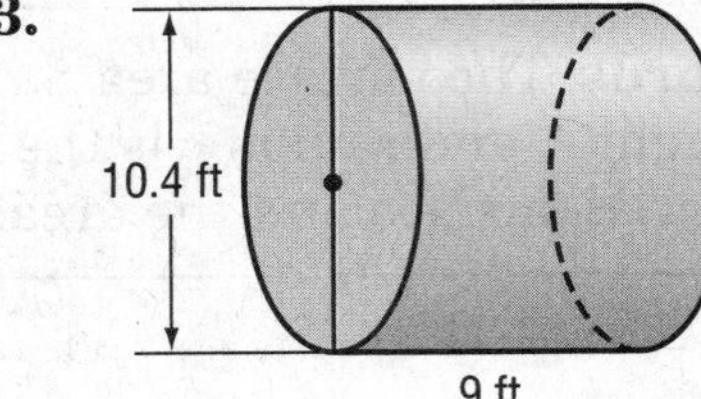

4.

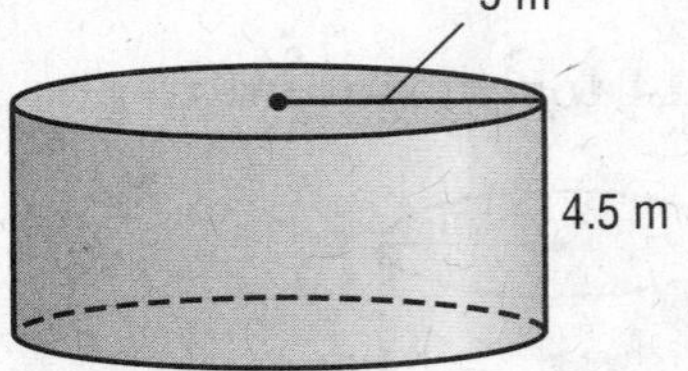

5.

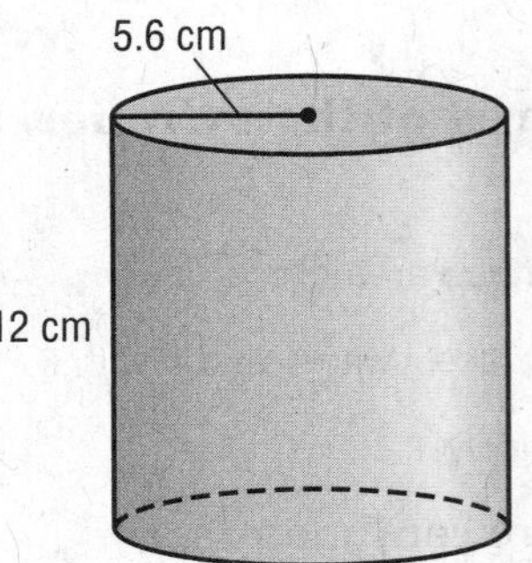

6.

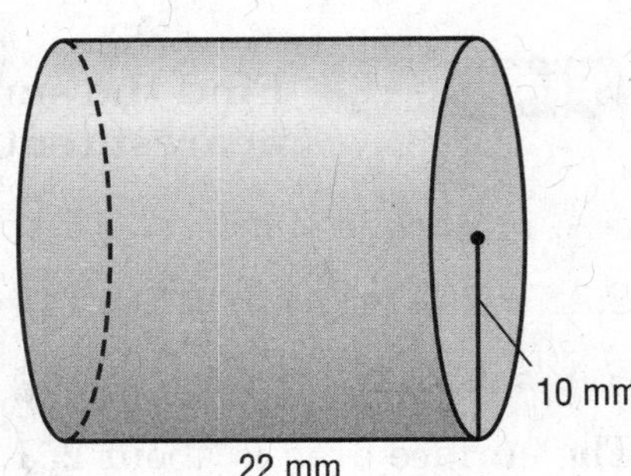

7.

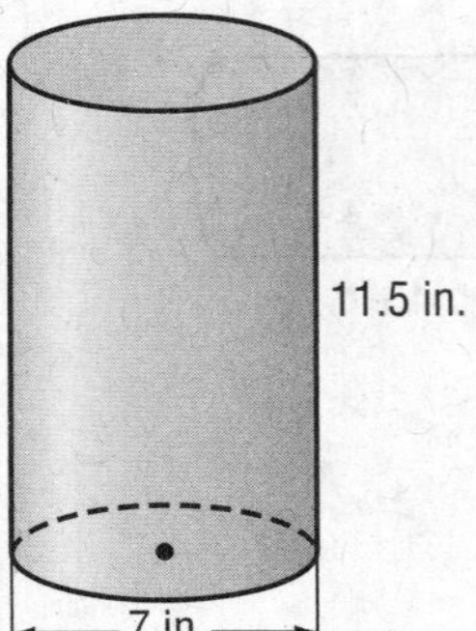

8.

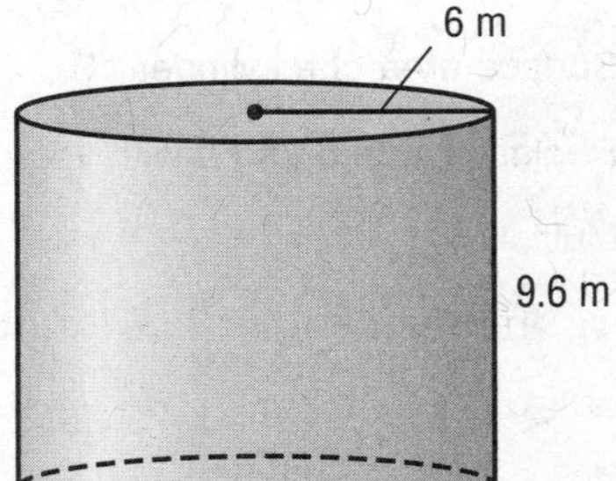

9.

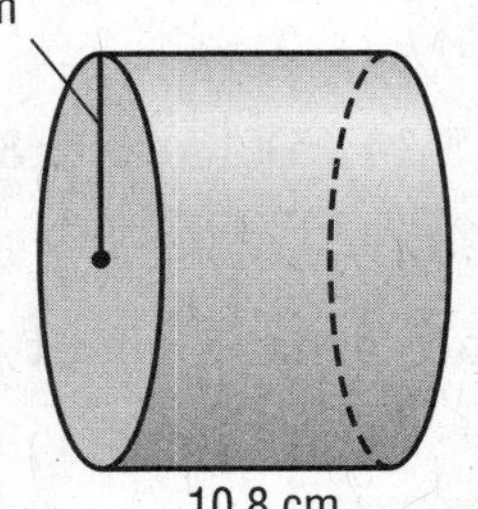

10. **CYLINDER:** radius, 9.4 mm; height, 15 mm

11. The net of a cylinder is shown. Find its surface area. Use 3.14 for π. Round to the nearest tenth.

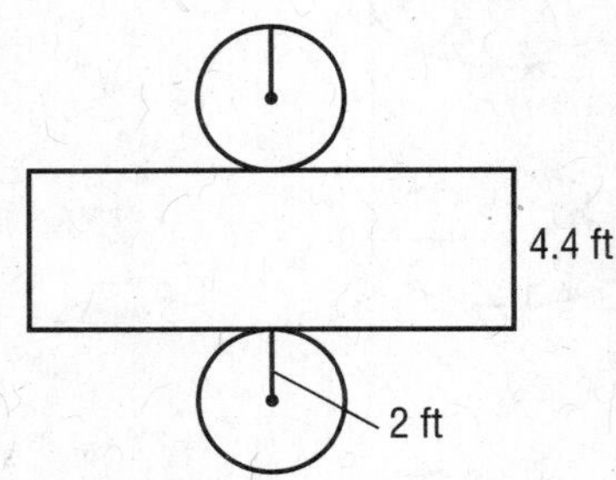

NAME ______________________ DATE __________ PERIOD ______

Reteach

Volume and Surface Area of Composite Figures

The volume or surface area of a composite figure can be found by separating the figure into solids with volumes or surface areas you know how to find.

Example 1 **Find the volume of the composite figure at the right.**

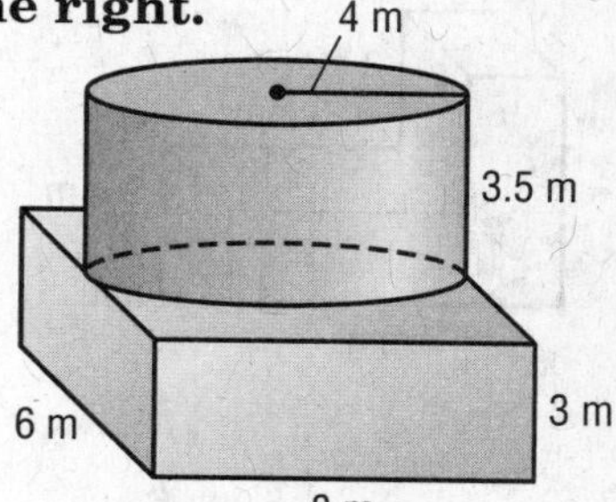

Find the volume of the prism and the cylinder.

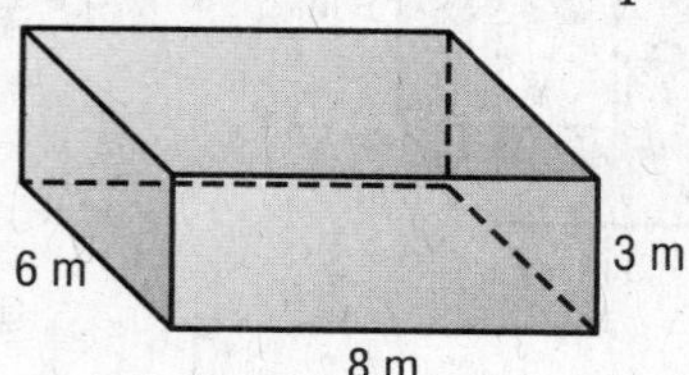

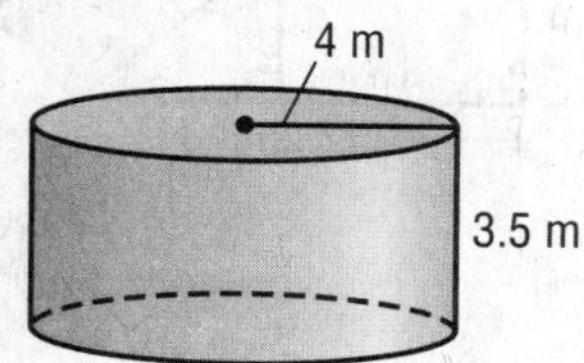

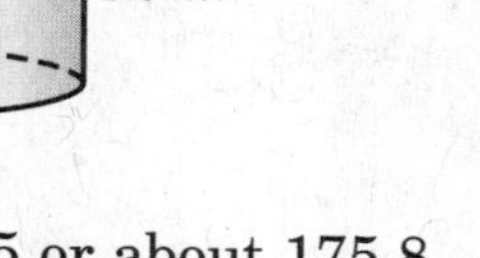

$V = bwh$
$V = 8 \cdot 6 \cdot 3$ or 144

$V = \pi r^2 h$
$V = \pi \cdot 4^2 \cdot 3.5$ or about 175.8

The volume of the composite shape is about 144 + 175.8, or 319.8 cubic meters.

Example 2 **Find the surface area of the figure above.**

Find the surface area of each shape. Then subtract twice the area of the base of the cylinder ($2\pi r^2 = 32\pi$, or about 100.5) since it is *not* a surface of the composite figure.

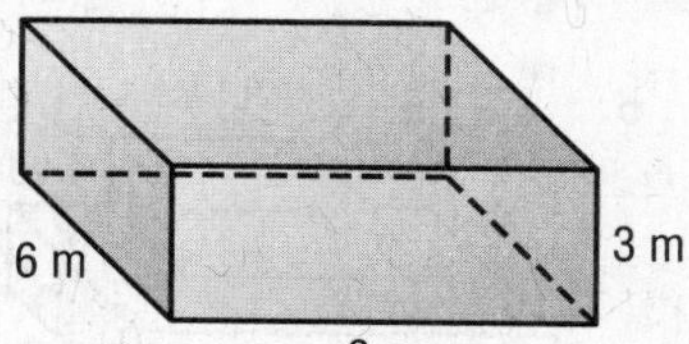

4 m
3.5 m

$S.A. = 2bh + 2bw + 2hw$
$S.A. = 2(8 \cdot 3) + 2(8 \cdot 6) + 2(3 \cdot 6)$
$S.A. = 48 + 96 + 36$
$S.A. = 180$

$S.A. = 2\pi rh + 2\pi r^2$
$S.A. = 2\pi(4)(3.5) + 2\pi(4)^2$
$S.A. \approx 87.92 + 100.48$
$S.A. \approx 188.4$

The surface area of the shape is about 180 + 188.4 – 100.5, or 267.9 square meters.

Exercises

1. Find the volume of the composite shape.

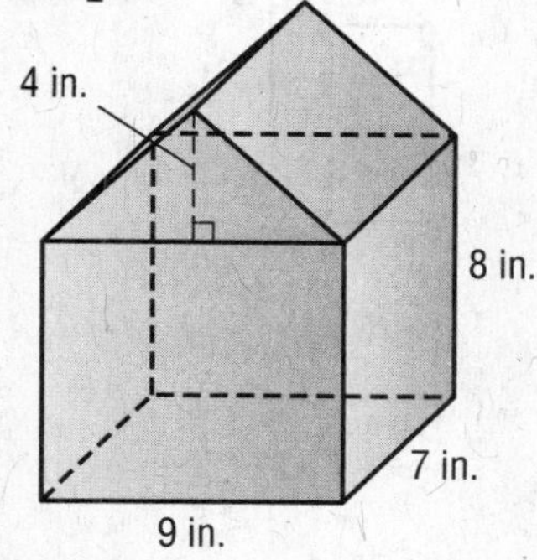

2. Find the surface area of the composite shape.

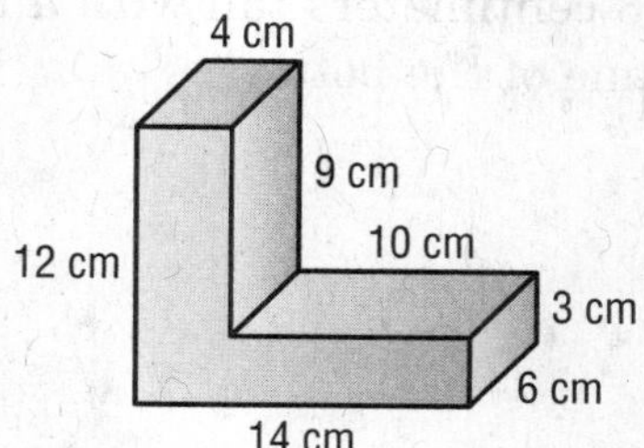

NAME ______________________ DATE ____________ PERIOD _____

Skills Practice

Volume and Surface Area of Composite Figures

Find the volume of each composite figure. Round to the nearest tenth if necessary.

1.

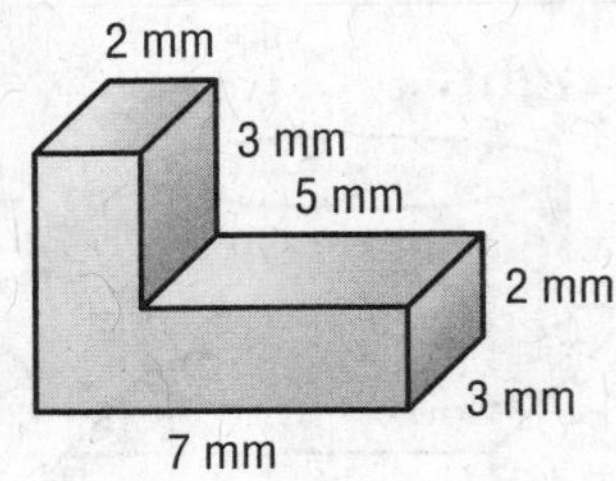

2.

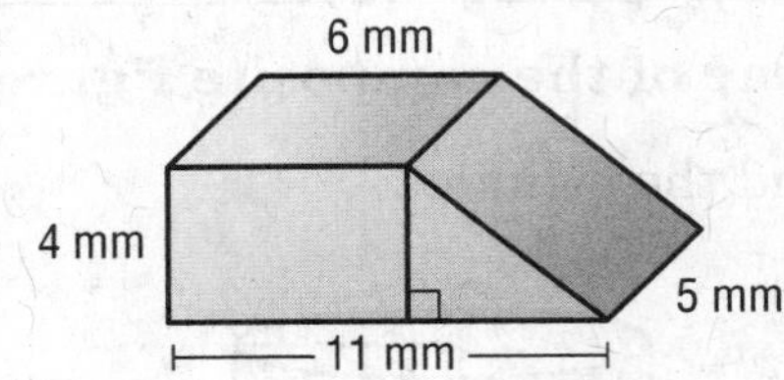

3.

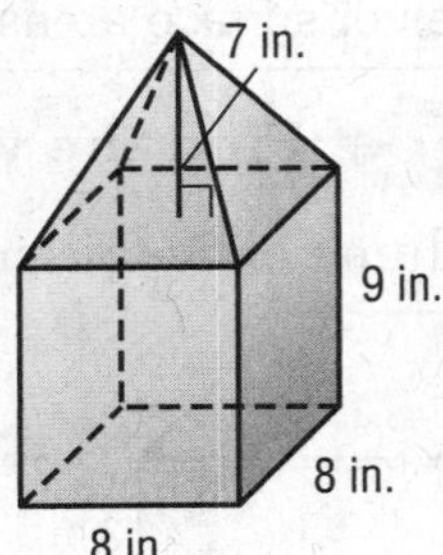

Find the surface area of each composite figure. Use 3.14 for π. Round to the nearest tenth if necessary.

4.

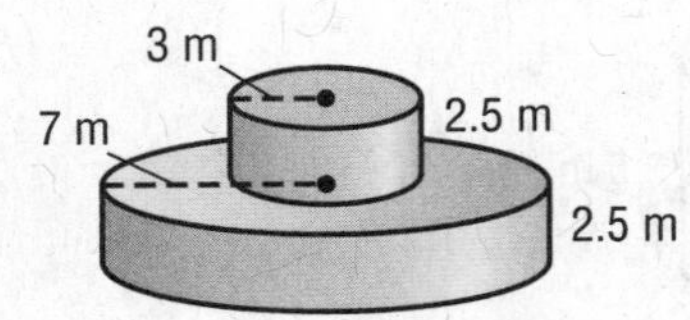

5.

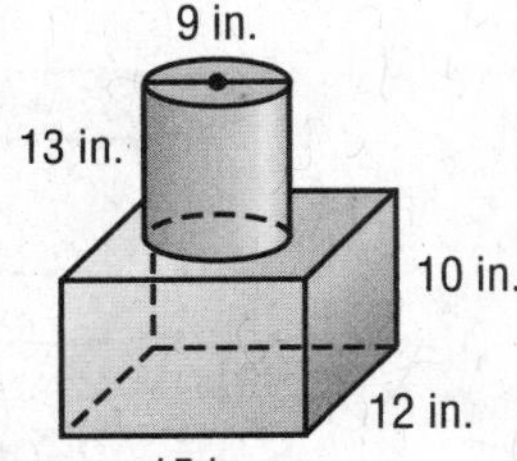

6.

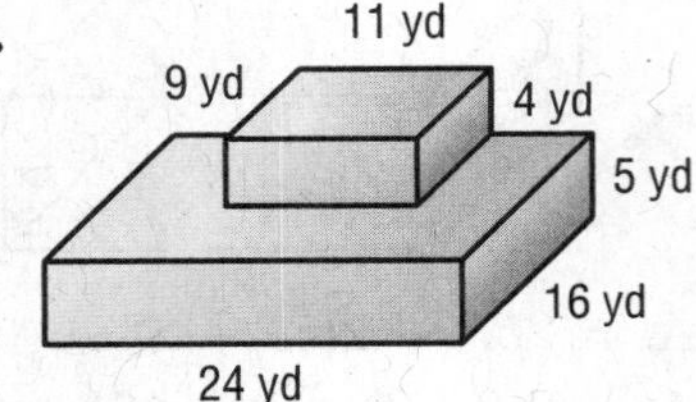

7. PERFUME A perfume bottle is shaped like a rectangular prism with a cylindrical neck, as shown at right. If the prism is 3 centimeters wide by 6 centimeters tall by 4 centimeters long, and the neck is 3 centimeters tall with a radius of 1 centimeter, what is the volume of the bottle?

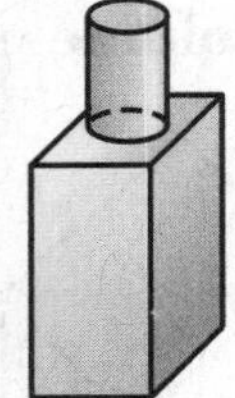

NAME ______________________ DATE ____________ PERIOD _____

Reteach

Mean

The **mean** of a data set is the sum of the data divided by the number of pieces of data.

Example 1 **The pictograph shows the number of members on four different swim teams. Find the mean number of members for the four different swim teams.**

Swim Team Members	
Amberly	
Carlton	
Hamilton	
West High	

Key: = 1 swimmer

$$\text{mean} = \frac{9 + 11 + 6 + 10}{4}$$

$$= \frac{36}{4} \text{ or } 9$$

Exercises

Find the mean for each set of data.

1.

Month	Snowfall (in.)
Nov.	20
Dec.	19
Jan.	20
Feb.	17
Mar.	4

2.

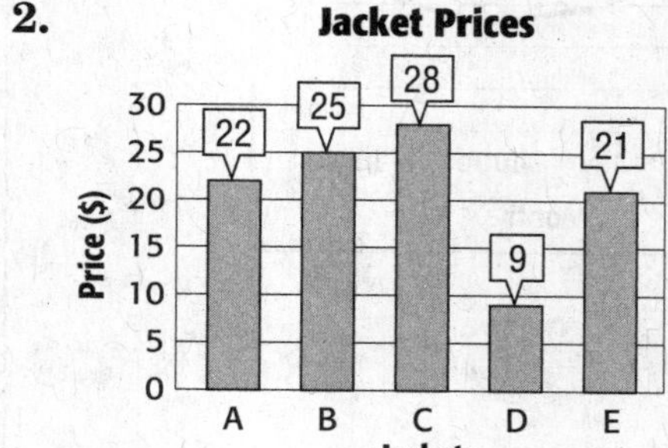

3.

Number of Bicycles	
Smiths	
Castros	
Lius	

Key: = 1 bicycle

4.

Checker Pieces	
A	
B	
C	
D	

Key: = 1 checker piece

NAME ______________________ DATE ____________ PERIOD _____

Skills Practice

Mean

Find the mean for each set of data.

1.

Number of Candy Bars Sold	
Amber	CANDY CANDY CANDY CANDY CANDY CANDY CANDY CANDY CANDY CANDY CANDY CANDY
Dalton	CANDY CANDY CANDY CANDY CANDY CANDY CANDY CANDY CANDY
Juan	CANDY CANDY CANDY CANDY CANDY CANDY CANDY CANDY CANDY CANDY
Shamika	CANDY CANDY CANDY CANDY CANDY CANDY CANDY CANDY CANDY CANDY CANDY CANDY CANDY

Key: CANDY = 1 candy bar

2.

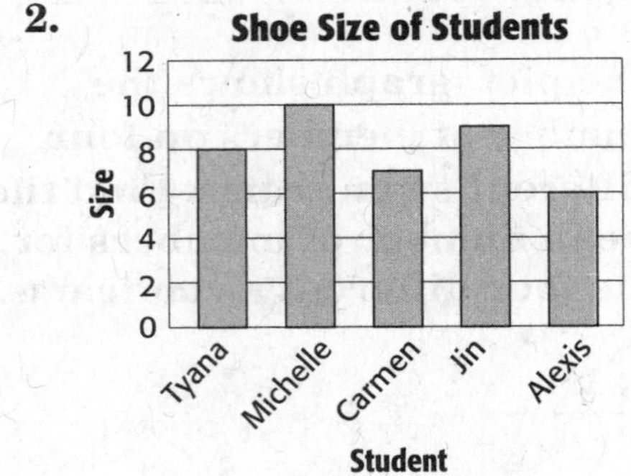

3.

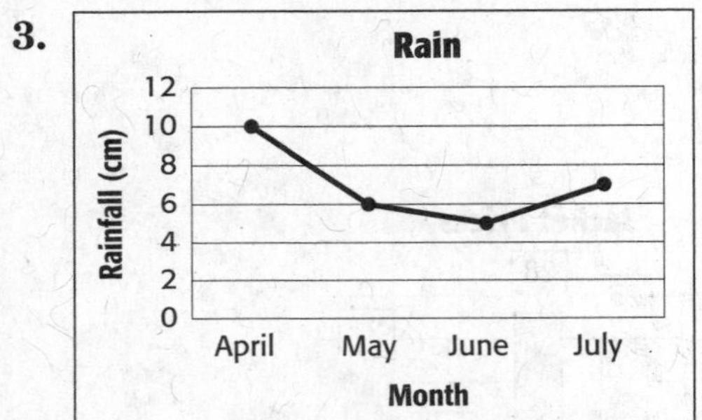

4.

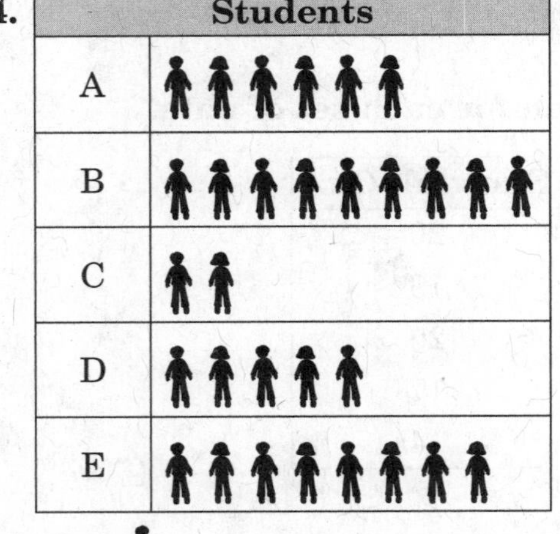

Key: = 1 student

5.

Temperatures	
Day	**Temp. (°F)**
Monday	69
Tuesday	70
Wednesday	73
Thursday	35
Friday	68

6.

Heights	
Student	**Height (in.)**
Maria	62
Peter	67
Shann	64
Iyoka	65
Evangelina	59
Carles	67

Multi-Part Lesson 1 PART D

NAME ______________________ DATE ____________ PERIOD ______

Reteach

Median, Mode, and Range

The **median** is the middle number of the data put in order, or the mean of the middle two numbers.
The **mode** is the number or numbers that occur most often.

Example 1 **The table shows the costs of seven different books. Find the mean, median, and mode of the data.**

Book Costs ($)			
22	13	11	16
14	13	16	

mean: $\frac{22 + 13 + 11 + 16 + 14 + 13 + 16}{7} = \frac{105}{7}$ or 15

To find the median, write the data in order from least to greatest.
median: 11, 13, 13, (14), 16, 16, 22

To find the mode, find the number or numbers that occur most often.
mode: 11, (13, 13), 14, (16, 16), 22

The mean is $15. The median is $14. There are two modes, $13 and $16.

The **range** of a set of data describes how the data vary.

Example 2 **Find the range of the data in the table. Then write a sentence describing how the data vary.**

Temperatures		
32°	40°	50°
55°	60°	63°

The greatest value is 63. The least value is 32. So, the range is 63° − 32° or 31°. The range is large. It tells us that the data vary greatly in value.

Exercises

Find the mean, median, mode, and range of each set of data.

1. hours worked: 14, 13, 14, 16, 8

2. points scored by a football team: 29, 31, 14, 21, 31, 22, 20

3.

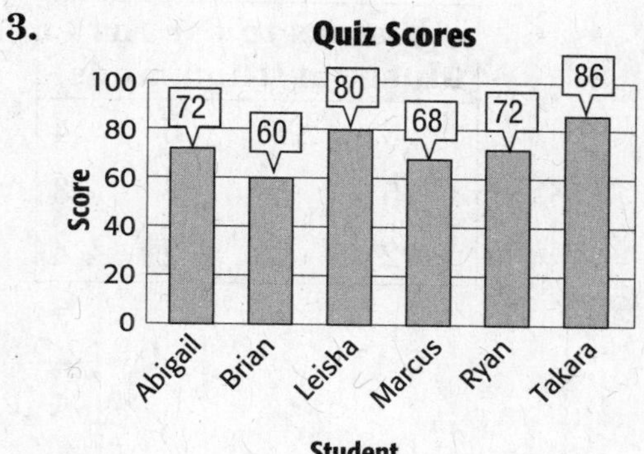

4.

Snowfall (in.)					
0	2	2	3	3	3
5	5	6	7	8	

NAME ______________________ DATE ____________ PERIOD ______

Skills Practice

Median, Mode, and Range

Find the median, mode, and range for each set of data.

1. age of children Danielle babysits: 6, 9, 2, 4, 3, 6, 5

2. hours spent studying: 13, 6, 7, 13, 6

3. age of grandchildren: 1, 15, 9, 12, 18, 9, 5, 14, 7

4. points scored in video game: 13, 7, 17, 19, 7, 15, 11, 7

5. amount of weekly allowances: 3, 9, 4, 3, 9, 4, 2, 3, 8

6. height of trees in feet: 25, 18, 14, 27, 25, 14, 18, 25, 23

Find the mean, median, mode, and range of the data represented.

7.

Annual Rainfall (in.)			
21	23	27	28
32	32	34	43

8.

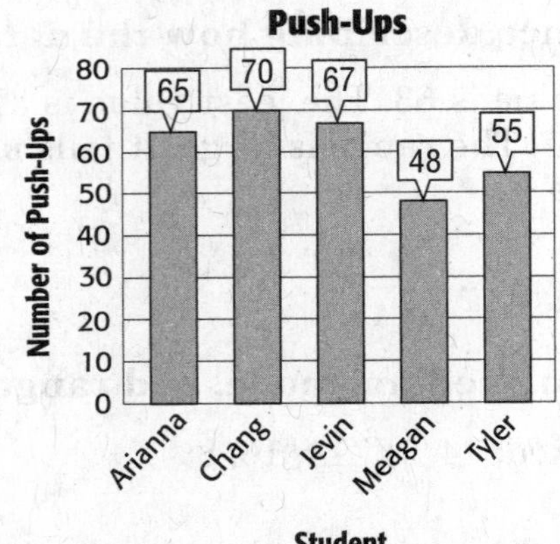

9. MUSEUMS Use the table showing the number of visitors to the art museum each month.

a. What is the mean of the data?

b. What is the median of the data?

c. What is the mode of the data?

Vistors to the Art Museum (thousands)			
3	11	5	4
5	3	6	3
12	2	2	4

NAME ______________________ DATE ____________ PERIOD _____

Multi-Part Lesson 2 PART A

Reteach

Frequency Tables

Example PETS **Mario asked his classmates how many pets they have. The results are shown in the table. Make a frequency table of the data.**

Number of Pets							
3	1	2	3	6	4	2	0
0	0	1	2	2	1	3	4
2	1	2	0	5	5	4	0

Make one tally in the frequency table for each time a particular number of pets occurs. Count and record the number of tallies.

Frequency Table

Number of Pets	Tally	Frequency
0	~~||||~~	5
1	||||	4
2	~~||||~~ |	6
3	|||	3
4	|||	3
5	||	2
6	|	1

Exercises

BASEBALL **The numbers of games won by the school baseball team over the last 15 years are shown below.**

10, 8, 11, 7, 9, 12, 13, 9, 7, 8, 10, 10, 9, 8, 8

1. Make a frequency table of the data.

Number of Wins	Tally	Frequency

2. How many times did the team win 10 or more games?

3. What fraction of the years did the baseball team win 9 games?

Multi-Part Lesson 2
PART A

NAME ______________________ DATE ____________ PERIOD _____

Skills Practice

Frequency Tables

ARTISTS The table shows the names of several famous artists.

Famous Artists			
Matisse	Monet	Cezanne	Picasso
Manet	Renoir	Rothko	Whistler
Dali	Van Gogh	Magritte	Degas
Miro	Da Vinci	Gauguin	Chagall

1. Make a frequency table to show the number of letters in each name.

2. Find the fraction of the names that have only 4 letters.

HOBBIES Violet took a survey of her classmates' hobbies. Her results appear in the table.

Classmates' Hobbies					
R	D	S	R	D	R
P	P	S	S	S	P
T	S	T	S	S	D
T	R	S	D	D	S

R = reading
D = drawing
P = photography
S = sports
T = watching TV

3. Make a frequency table of the data.

4. What fraction of the classmates chose photography?

NAME ______________________ DATE __________ PERIOD _____

Reteach

Circle Graphs

Example 1 **Create a circle graph to display the data shown at the right.**

Pies in a Baking Contest	
Pie	**Percent**
Apple	52
Cherry	24
Peach	12
Pecan	12

Step 1 Write a fraction to estimate each percent.

$52\% \approx 50\%$ and $50\% = \frac{50}{100}$ or $\frac{1}{2}$

$24\% \approx 25\%$ and $25\% = \frac{25}{100}$ or $\frac{1}{4}$

$12\% \approx 10\%$ and $10\% = \frac{10}{100}$ or $\frac{1}{10}$

Step 2 Use a compass to draw a circle with at least a 1-inch radius.

Step 3 Since 52% is a little more than $\frac{1}{2}$, divide a little more than $\frac{1}{2}$ of the circle for "apple." Since 24% is a little less than $\frac{1}{4}$, divide a little less than $\frac{1}{4}$ of the circle for "cherry." Since the last two sections are equal, take the remaining portion of the circle and divide it into two equal parts.

Step 4 Label each section of the circle graph. Give the graph a title.

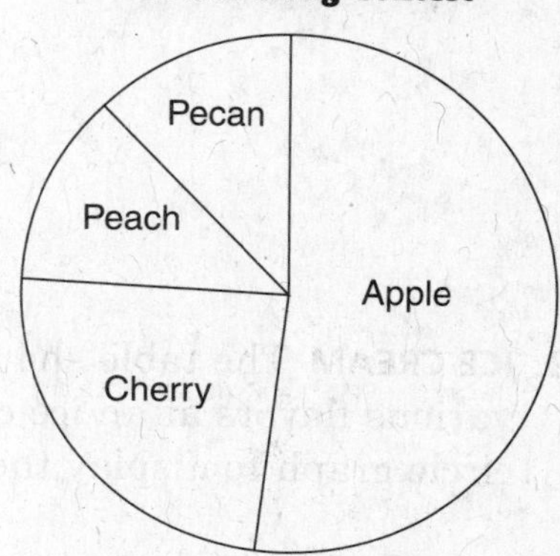

Example 2 **BEVERAGES Jesse surveyed his class to see what type of beverage they drink most often. The circle graph shows his results. Suppose Jesse asks 300 students in his school which beverage they drink most often. About how many would say water?**

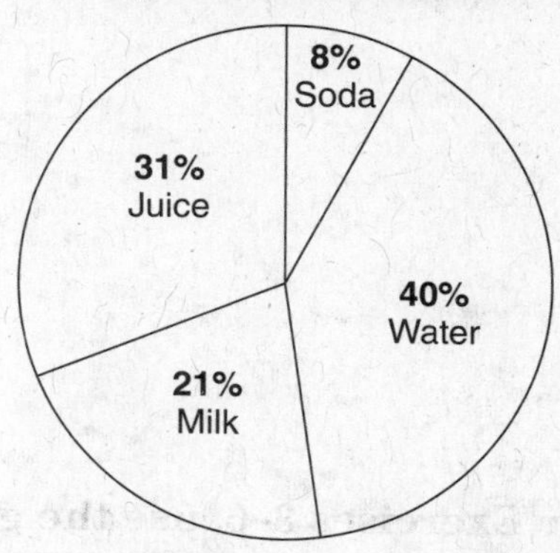

Find 40% of 300 or 0.4×300. — Write 40% as a decimal.

$0.4 \times 300 = 120$ — Multiply.

So, about 120 students would say they drink water most often.

Exercises

1. **SCHOOL** Parents were asked whether they were in favor of year-round school. Thirty-four percent responded "yes", 58 percent responded "no", and 8 percent responded "unsure". Create a circle graph to represent the data.

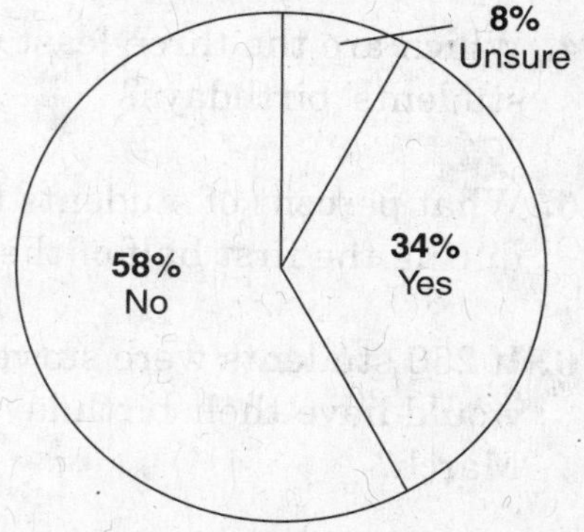

2. Refer to the graph in Example 2 above. Which beverages together are drunk most often by about half of the students?

NAME ______________________ DATE ____________ PERIOD _____

Skills Practice

Circle Graphs

1. **MASCOT** The table shows the results of how students voted for the new school mascot. Create a circle graph to display the data.

Votes for School Mascot	
Mascot	**Percent**
Bear	12
Eagle	71
Lion	10
Tiger	7

2. **ICE CREAM** The table shows the daily sales of various flavors at an ice cream shop. Create a circle graph to display the data.

Ice Cream Sales	
Flavor	**Percent**
Vanilla	4
Chocolate	19
Strawberry	4
Cookies & Cream	34
Butter Pecan	25
Peanut Butter Cup	14

For Exercises 3–6, use the graph below that shows the months of students' birthdays.

3. Which three months have the most birthdays?

4. Which are the three least common months for students' birthdays?

5. What percent of students have their birthday during the first half of the year?

6. If 250 students were surveyed, about how many would have their birthday during January through March?

Months of Student's Birthdays

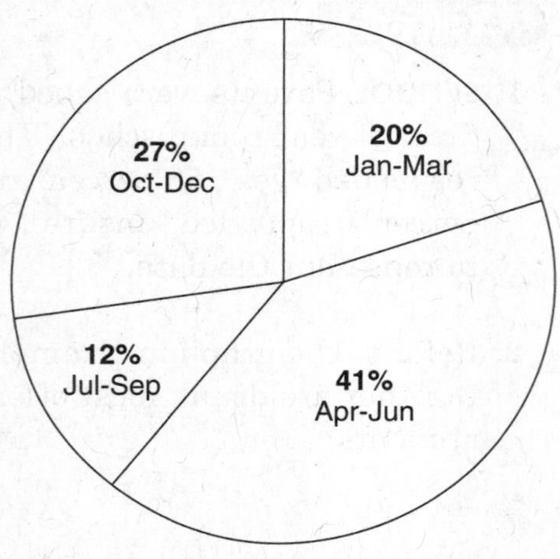

NAME ______________________ DATE ____________ PERIOD ______

Reteach

Select an Appropriate Display

Data can be displayed in many different ways, including the following:

- A **bar graph** shows the number of items in a specific category.
- A **line graph** shows change over a period of time.
- A **line plot** shows how many times each number occurs in the data.
- A **frequency table** shows the number of times each data value appears.
- A **pictograph** uses picture symbols to compare data.

Example 1 **Which display allows you to see how art show ticket prices have changed since 2007?**

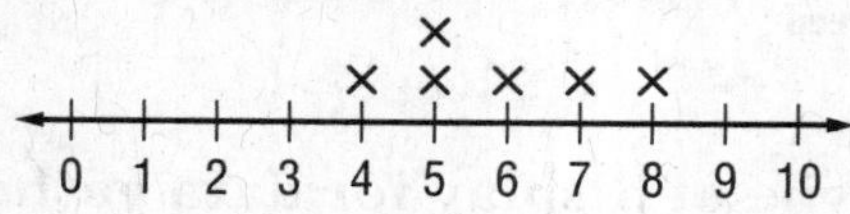

The line graph allows you to see how the art show ticket prices have increased since 2007.

Example 2 **What type of display would you use to show the results of a survey of students' favorite brand of tennis shoes?**

Since the data would list the number of students that chose each brand, or category, the data would best be displayed in a bar graph.

Exercises

1. **GRADES** Which display makes it easier to see how many students had test scores in the 80s?

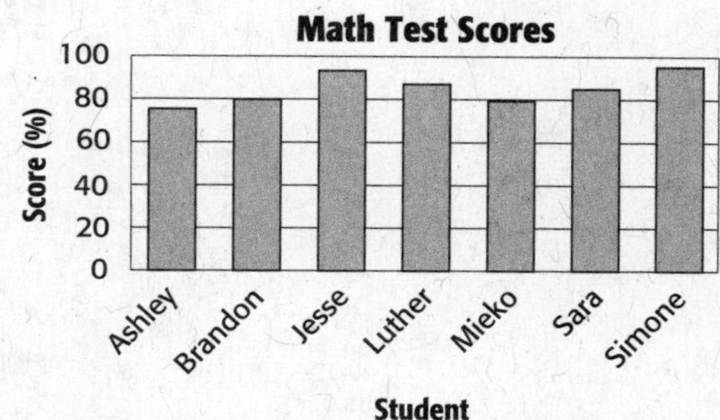

Math Test Scores	
Scores	**Number of Students**
70–79	2
80–89	3
90–100	2

2. **VOLLEYBALL** What type of display would you use to show the number of wins the school volleyball team had from 2000 to 2005?

NAME ______________________ DATE __________ PERIOD ______

Skills Practice

Select an Appropriate Display

1. **ANIMALS** Which display makes it easier to compare the average weight of a bulldog with the average weight of a pug?

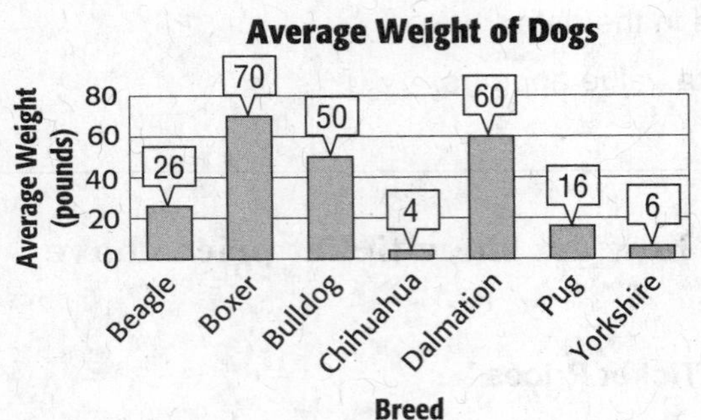

Average Weight of Dogs	
Weights	Number of Dogs
0–9	2
10–19	1
20–29	1
30–39	0
40–49	0
50–59	1
60–69	1
70–79	1

Select an appropriate type of display for data gathered about each situation.

2. record high temperature for each month this year

3. test scores each student had on a science test

4. favorite topping on a pizza of the students in Mrs. Witsken's class

5. Edmund's weight over the past 10 years

6. Select and make an appropriate type of display.

Company Sales	
Year	Sales (millions $)
2007	4.0
2008	4.5
2009	4.0
2010	5.5
2011	6.0
2012	8.0

Multi-Part Lesson 4 PART C

NAME ______________________ DATE ____________ PERIOD _____

Reteach

Misleading Graphs and Statistics

Graphs can be misleading for many reasons: there is no title, the scale does not include 0; there are no labels on either axis; the intervals on a scale are not equal; or the size of the graphics misrepresents the data.

Example

WEEKLY CHORES The line graphs below show the total hours Salomon spent doing his chores one month. Which graph would be best to use to convince his parents he deserves a raise in his allowance? Explain.

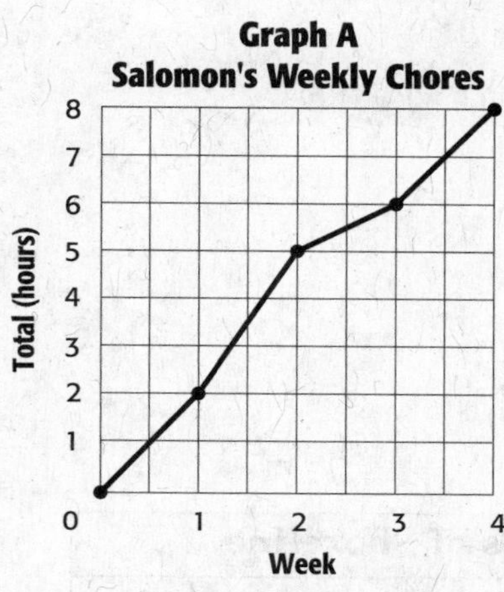

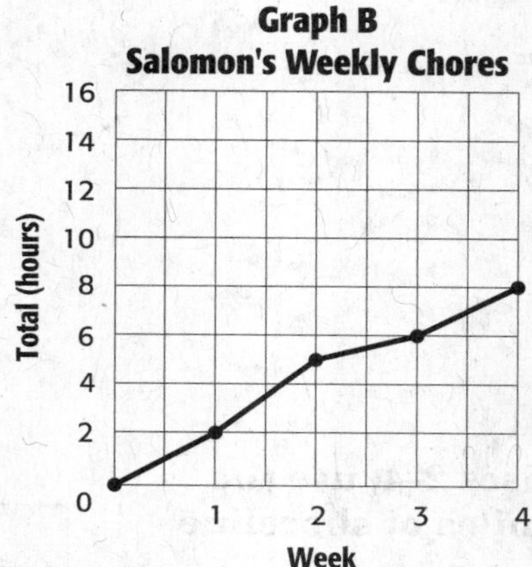

He should use graph A because it makes the total hours seem much larger.

Exercises

PROFITS For Exercises 1 and 2, use the graphs below. It shows a company's profits over a four-month period.

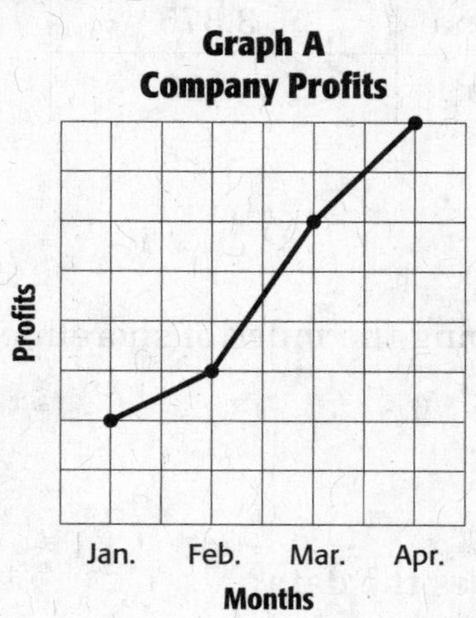

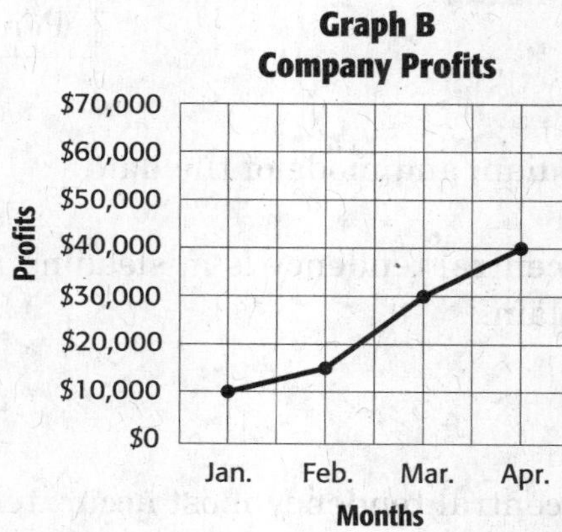

1. Which graph would be best to use to convince potential investors to invest in this company?

2. Why might the graph be misleading?

NAME ______________________ DATE ____________ PERIOD _____

Skills Practice

Misleading Graphs and Statistics

1. LUNCH Which graph could be used to indicate a greater increase in yearly lunch prices? Explain.

Lunch Prices

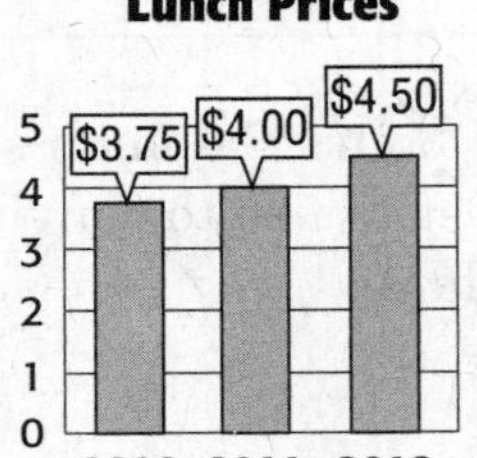

Graph A

Lunch Prices ($)

Graph B

GEOGRAPHY For Exercises 2–4, use the table that shows the miles of shoreline for five states.

Miles of Shoreline	
State	Length of Shoreline (mi)
Virginia	3,315
Maryland	3,190
Washington	3,026
North Carolina	3,375
Pennsylvania	89

2. Find the mean, median, and mode of the data.

3. Which measure of central tendency is misleading in describing the miles of shoreline for the states? Explain.

4. Which measure of central tendency most accurately describes the data?

NAME ______________________ DATE __________ PERIOD ______

Reteach

Problem-Solving Investigation: Use Logical Reasoning

You may need to use a Venn diagram to solve some problems.	
Example	• Determine what information is given in the problem and what you need to find.
Plan	• Select a strategy including a possible estimate.
Solve	• Solve the problem by carrying out your plan.
Check	• Examine your answer to see if it seems reasonable.

Of the 25 skiers on the ski team, 13 signed up to race in the Slalom race, and 8 signed up for the Giant Slalom race. Six skiers signed up to ski in both the Slalom and the Giant Slalom races. How many skiers did not sign up for any races?

Understand You know how many skiers signed up for each race and how many signed up for both races. You need to organize the information.

Plan You can use a Venn diagram to organize the information.

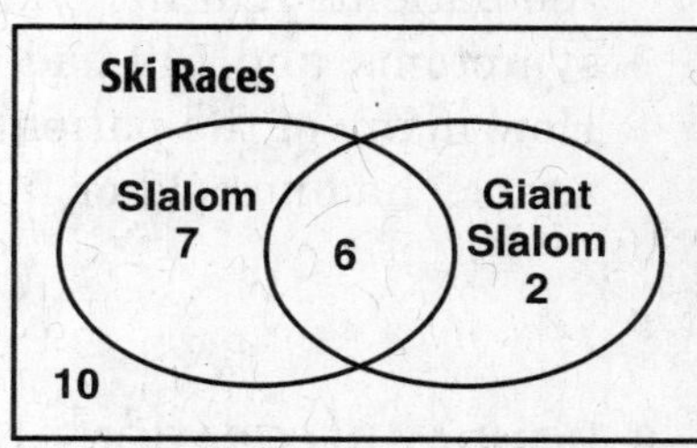

Solve Draw two overlapping circles to represent the two different races. Place a 6 in the section that is a part of both circles. Use subtraction to determine the number for each other section.

only the Slalom race: $13 - 6 = 7$

only the Giant Slalom race: $8 - 6 = 2$

neither the Slalom or the Giant Slalom race:

$25 - 7 - 2 - 6 = 10$

There were 10 skiers who did not sign up for either race.

Check Check each circle to see if the appropriate number of students is represented.

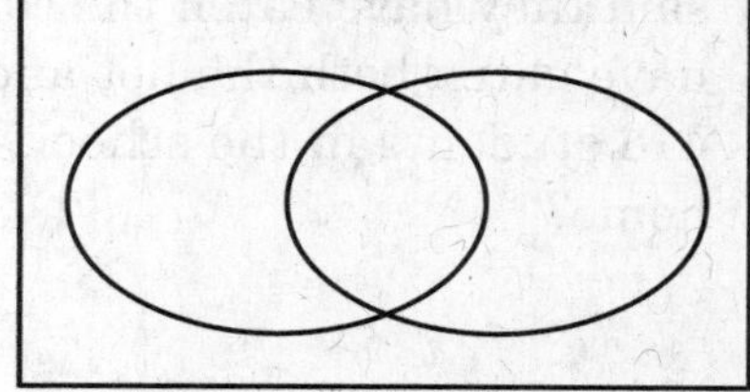

Exercise

NAME ______________________ DATE ____________ PERIOD _____

Skills Practice

Problem-Solving Investigation: Use Logical Reasoning

Use a Venn diagram to solve each problem.

1. **PHONE SERVICE** Of the 5,750 residents of Homer, Alaska, 2,330 pay for landline phone service and 4,180 pay for cell phone service. One thousand seven hundred fifty pay for both landline and cell phone service. How many residents of Homer do not pay for any type of phone service?

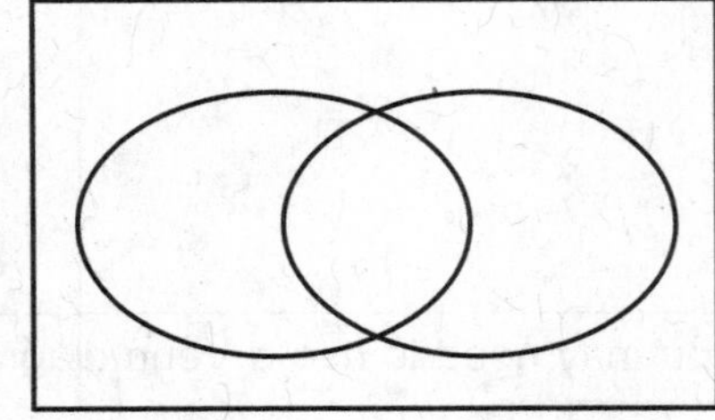

2. **BIOLOGY** Of the 2,890 ducks living in a particular wetland area, scientists find that 1,260 have deformed beaks, while 1,320 have deformed feet. Six hundred ninety of the birds have both deformed feet and beaks. How many of the ducks living in the wetland area have no deformities?

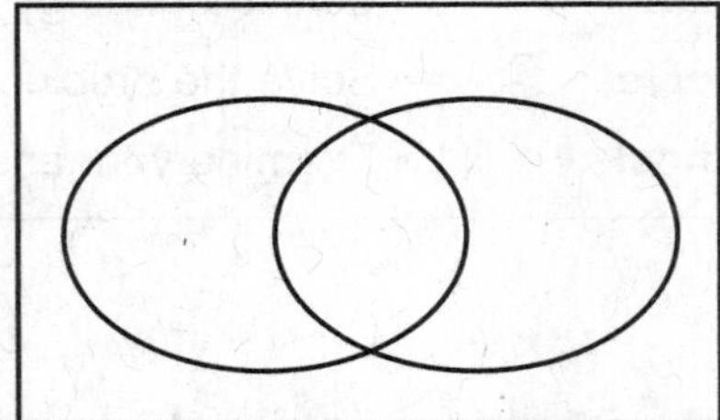

3. **FLU SYMPTOMS** The local health agency treated 890 people during the flu season. Three hundred fifty of the patients had flu symptoms, 530 had cold symptoms, and 140 had both cold and flu symptoms. How many of the patients treated by the health agency had no cold or flu symptoms?

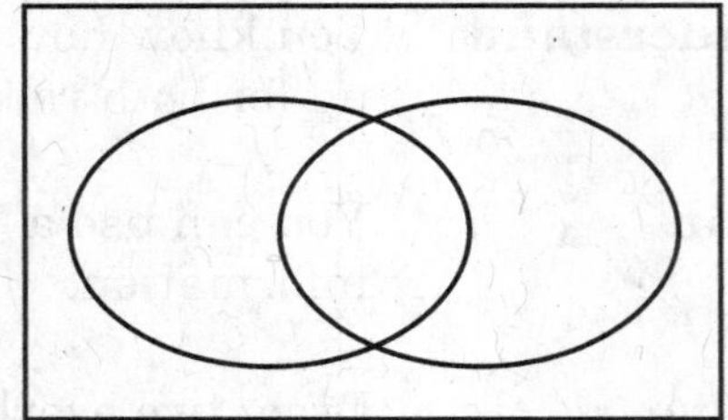

4. **HOLIDAY DECORATIONS** During the holiday season, 13 homes on a certain street displayed lights and 8 displayed lawn ornaments. Five of the homes displayed both lights and lawn ornaments. If there are 32 homes on the street, how many had no decorations at all?

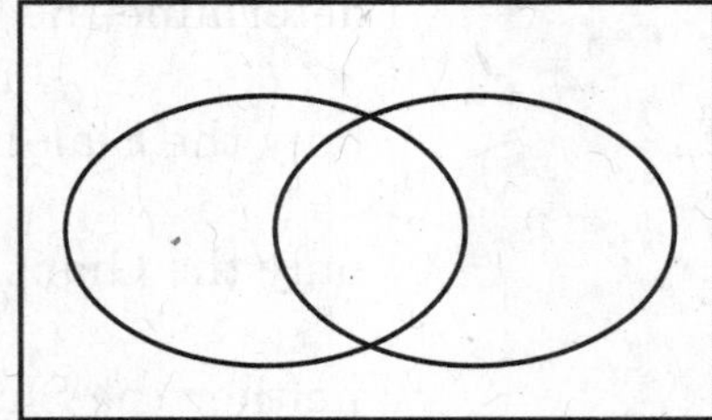

5. **LUNCHTIME** At the local high school, 240 students reported they have eaten the cafeteria's hot lunch, 135 said they have eaten the cold lunch, and 82 said they have eaten both the hot and cold lunch. If there are 418 students in the school, how many bring lunch from home?

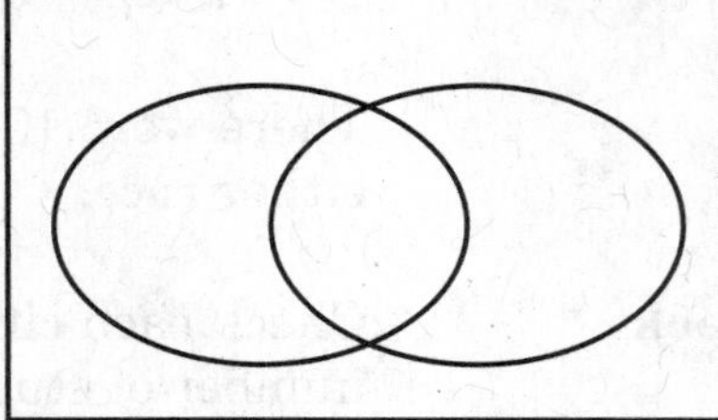

NAME ________________________ DATE ____________ PERIOD _____

Reteach

Probability of Simple Events

When tossing a coin, there are two possible **outcomes**, heads and tails. Suppose you are looking for heads. If the coin lands on heads, this would be a favorable outcome. The chance that some event will happen (in this case, getting heads) is called **probability**. You can use a ratio to find probability. The probability of an event is a number from 0 to 1, including 0 and 1. The closer a probability is to 1, the more likely it is to happen.

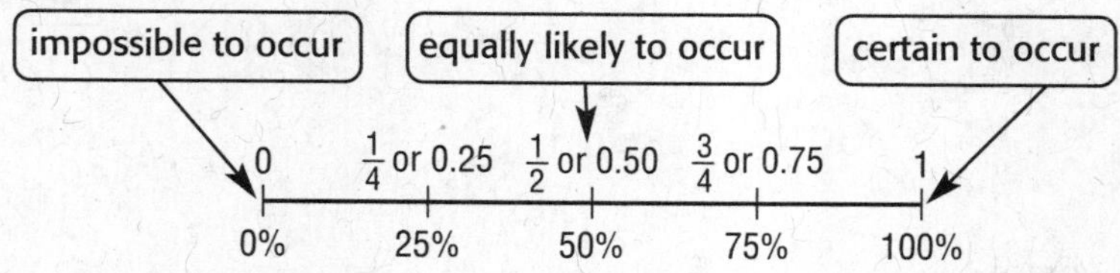

Example 1 **There are four equally likely outcomes on the spinner. Find the probability of spinning green or blue.**

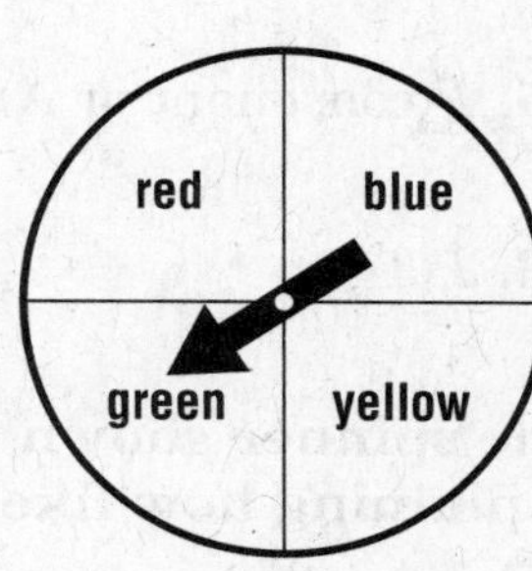

$$P(\text{green or blue}) = \frac{\text{number of favorable outcomes}}{\text{number of total outcomes}}$$

$$= \frac{2}{4} \text{ or } \frac{1}{2}$$

The probability of landing on green or blue is $\frac{1}{2}$, 0.50, or 50%.

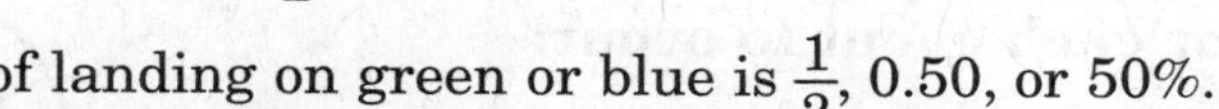

Complementary events are two events in which either one or the other must happen, but both cannot happen at the same time. The sum of the probabilities of complementary events is 1.

Example 2 **There is a 25% chance that Sam will win a prize. What is the probability that Sam will not win a prize?**

$$\begin{aligned} P(\text{win}) + P(\text{not win}) &= 1 \\ 0.25 + P(\text{not win}) &= 1 \\ -0.25 \qquad\qquad &= -0.25 \\ \hline P(\text{not win}) &= 0.75 \end{aligned}$$

So, the probability that Sam won't win a prize is 0.75, 75%, or $\frac{3}{4}$.

Exercises

1. There is a 90% chance that it will rain. What is the probability that it will not rain?

One pen is chosen without looking from a bag that has 3 blue pens, 6 red, and 3 green. Find the probability of each event. Write each answer as a fraction, a decimal, and a percent.

2. P(green)
3. P(blue or red)
4. P(*not* red)

NAME ______________________ DATE ____________ PERIOD _____

Skills Practice

Probability of Simple Events

A card is randomly chosen. Find each probability. Write each answer as a fraction, a decimal, and a percent.

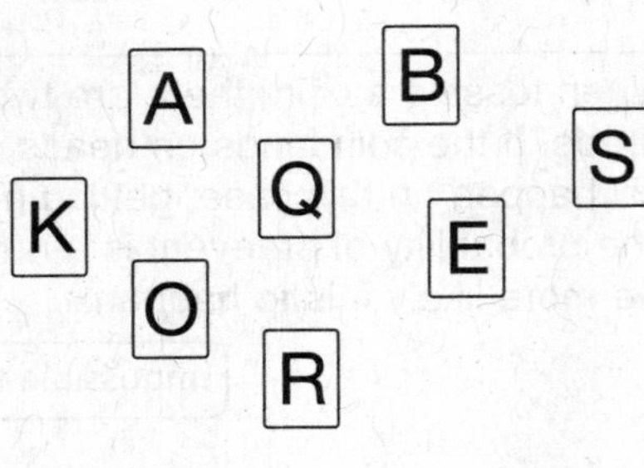

1. *P*(B)

2. *P*(Q or R)

3. *P*(vowel)

4. *P*(consonant or vowel)

5. *P*(consonant or A)

6. *P*(T)

The spinner shown is spun once. Write a sentence explaining how likely it is for each event to occur.

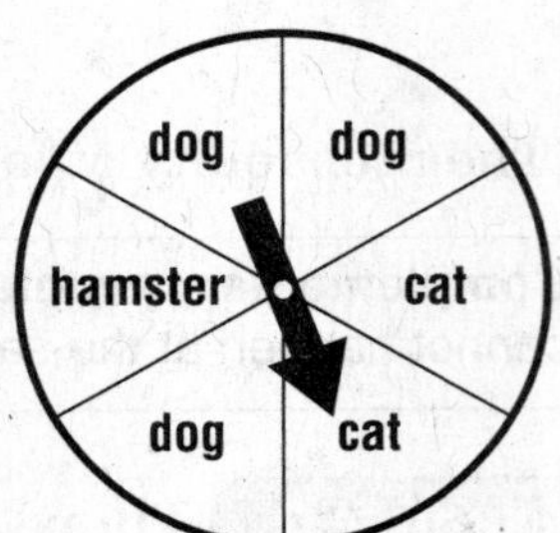

7. *P*(dog)

8. *P*(hamster)

9. *P*(dog or cat)

10. *P*(bird)

11. *P*(mammal)

WEATHER The weather reporter says that there is a 12% chance that it will be moderately windy tomorrow.

12. What is the probability that it will not be windy?

13. Will tomorrow be a good day to fly a kite? Explain.

NAME ______________________ DATE ____________ PERIOD _____

Reteach

Sample Spaces

You can use a tree diagram or list to find the number of possible outcomes.

Example 1 **How many sandwiches are possible from a choice of turkey or ham with jack cheese or Swiss cheese?**

Draw a tree diagram.

Sandwich	Cheese	Outcome
turkey (T)	jack (J)	TJ
	Swiss (S)	TS
ham (H)	jack (J)	HJ
	Swiss (S)	HS

There are four possible sandwiches.

Example 2 **Use a list to find, how many sandwiches are possible from a choice of roast beef, turkey, or ham, with a choice of jack, cheddar, American, or Swiss cheese. Find the probability of choosing a ham sandwich with jack cheese.**

roast beef, jack	turkey, jack	ham, jack
roast beef, cheddar	turkey, cheddar	ham, cheddar
roast beef, American	turkey, American	ham, American
roast beef, Swiss	turkey, Swiss	ham, Swiss

There are 12 possible outcomes. Only one outcome is a ham sandwich with jack cheese. So, the probability is $\frac{1}{12}$.

Exercises

Use a tree diagram or list to determine the number of possible outcomes. Then find the probability.

1. Buy a can or a bottle of grape or orange soda
   Find *P*(bottle, grape).

2. Toss a coin and roll a number cube
   Find *P*(4, tails).

3. wear jeans or shorts with a blue, white, black, or red T-shirt. Find *P*(jeans, white T-shirt).

Skills Practice

Sample Spaces

1. In how many ways can 2 coins be chosen from a set of 1 penny, 1 nickel, 1 dime, and 1 quarter? Make an organized list to show the sample space.

Use a tree diagram or list to determine the number of possible outcomes for each situation. Then find the probability.

2. Each spinner is spun once. How many outcomes are possible? Find *P*(pink, Z).

3. Chocolate, vanilla, strawberry, or mint ice cream with sugar or waffle cone. How many outcomes are possible? Find *P*(vanilla, waffle).

4. Paint room cream, violet, or blue with red, white or gold trim. How many outcomes are possible? Find *P*(blue, red).

NAME ______________________ DATE __________ PERIOD _____

Reteach

The Fundamental Counting Principle

The **Fundamental Counting Principle** states that if there are *m* outcomes for a first choice and *n* outcomes for a second choice, then the total number of possible outcomes can be found by multiplying $m \times n$.

Example 1 **A spinner with three equal sections marked A, B, and C is spun and a letter is randomly chosen from the word *flower*. Find the total number of possible outcomes.**

spinner choices		letter choices		total number of outcomes	
3	×	6	=	18	Fundamental Counting Principle

There are 18 possible outcomes.

Example 2 **A local deli offers customers a choice of three different kinds of soups, four different kinds of sandwiches, and two different beverages. How many meals consisting of one soup, one sandwich, and one beverage are possible?**

soup choices		sandwich choices		beverage choices		total number of outcomes	
3	×	4	×	2	=	24	Fundamental Counting Principle

There are 24 possible meals.

Exercises

Use the Fundamental Counting Principle to find the total number of possible outcomes in each situation.

1. selecting one letter from the word *tiger* and one month of the year

2. tossing a coin and rolling a number cube

3. selecting either Joseph or Yolanda to be president and either Beckie, Antwon, or Guido to be vice-president

4. selecting one entrée from chicken, beef, or fish and one side dish from salad, soup, or potato

5. spinning a spinner with eight equal sections twice

6. choosing a color from ten colors and a number from 1 to 5

7. Michelle has four pairs of pants and seven shirts. Assuming she can wear any of the shirts with each pair of pants, how many outfits are possible?

8. A spinner with five equal sections marked A, B, C, D, and E is spun twice. What is the probability that the spinner will land on A after the first spin and on C after the second spin?

NAME ______________________ DATE __________ PERIOD ______

Multi-Part Lesson 1 PART E

Skills Practice

The Fundamental Counting Principle

Use the Fundamental Counting Principle to find the total number of outcomes in each situation.

1 selecting one dessert from cake, pie, or ice cream and one beverage from milk, tea, juice, water, or soda

2. spinning the spinner shown three times

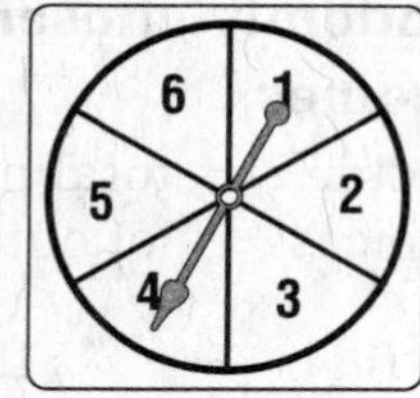

3. tossing a quarter and tossing a penny

4. selecting either Micah, Caleb, or Raheem as team captain and either Juan or Jacob as co-captain

5. selecting one letter from the word *excellent* and one day of the week

6. choosing a shape from seven shapes and a number from 1 to 10

7. choosing a pencil from twelve pencils and an eraser from three erasers

8. choosing a flavor from nine flavors and a topping from five toppings

9. choosing a letter from twenty letters and a number from 1 to 5

10. choosing a student from six students and a number from 1 to 8

11. Students have a choice of pizza, a hot dog, or macaroni and cheese for lunch. What is the probability that the first student in line will choose pizza and the next student in line will choose macaroni and cheese?

12. A number cube is rolled twice. What is the probability that the number cube will land on 5 after the first roll and on 2 after the second roll?

13. A theatre is showing eight different movies. What is the probability that two customers chosen at random will select the same movie?

14. Ava is surveying her classmates to determine their favorite season of the year. What is the probability that the first student surveyed will choose fall, and the next student surveyed will choose summer?

Multi-Part Lesson 2 PART B

NAME ______________________ DATE ____________ PERIOD ______

Reteach

Probability of Independent Events

Compound events in which the outcome of one event does not affect the outcome of the other event are **independent events**. The probability of two independent events can be found by multiplying the probability of the first event by the probability of the second event. Therefore, $P(A \text{ and } B) = P(A) \times P(B)$.

Example 1 **A coin is tossed and the spinner is spun. Find the probability of tossing tails and spinning a number greater than 2.**

First, find the probability of each event.

$P(\text{tails}) = \frac{1}{2}$ ⟵ number of ways to toss tails / number of possible outcomes

$P(\text{greater than 2}) = \frac{4}{6}$ ⟵ numbers greater than 2 / number of possible outcomes

Then find the probability of both events occurring.

$P(\text{tails and greater than 2}) = \frac{1}{2} \times \frac{4}{6}$ $P(\text{tails}) \times P(\text{greater than 2})$

$= \frac{4}{12} = \frac{1}{3}$ Multiply and simplify.

The probability of tossing tails and spinning greater than two is $\frac{1}{3}$.

Example 2 **Sara has 4 DVDs and 12 CDs in a box. She chose 1 item without looking, replaced it, and then chose a second item from the box. What is the probability that each item selected was a DVD?**

Find the probability that the first item selected is a DVD.

$P(\text{DVD}) = \frac{4}{16}$ or $\frac{1}{4}$

Since Sara replaces the item back into the box after the first trial, the probability that the second item selected is a DVD is also $\frac{4}{16}$ or $\frac{1}{4}$.

$P(\text{DVD and DVD}) = \frac{1}{4} \times \frac{1}{4}$ $P(\text{DVD}) \times P(\text{DVD})$

$= \frac{1}{16}$ Multiply.

So, the probability of selecting a DVD and a DVD is $\frac{1}{16}$.

Exercises

For Exercises 1–3, refer to the spinner from Example 1. Find each probability.

1. *P*(odd and even)
2. *P*(3 and 5)
3. *P*(1 and greater than 1)

4. Kaden has 2 pieces of gum, 3 mints, and 5 hard candies in a jar. He pulls out one item, then replaces it and pulls out another item. What is the probability that each item is a piece of gum? Write your answer as a fraction, decimal, and percent.

NAME ______________________ DATE ____________ PERIOD ____

Skills Practice

Probability of Independent Events

Two bags each contain a different set of cards. A card from Bag 1 is chosen, and a card from Bag 2 is chosen. Find the probability of each event.

Bag 1

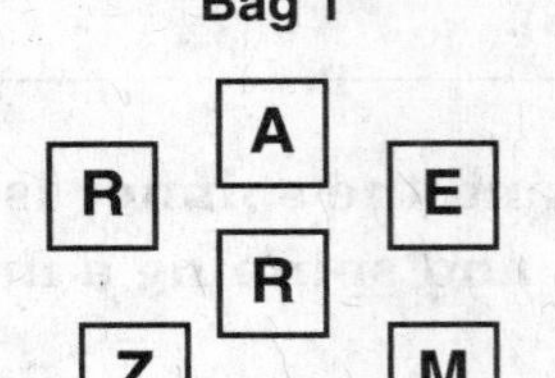

Bag 2

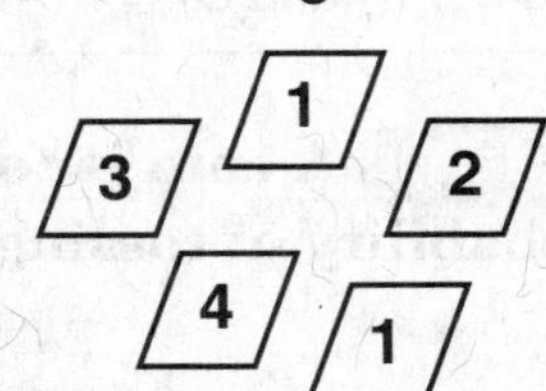

1. *P*(Z and 1)

2. *P*(R and less than 3)

3. *P*(vowel and 2)

4. *P*(*not* M and even)

5. *P*(E and *not* 4)

A rack contains one yellow kickball, four blue kickballs, and three red kickballs. One kickball is chosen from the rack without looking. Then it is replaced and another kickball is chosen. Find each probability.

6. *P*(red and blue)

7. *P*(yellow and yellow)

8. *P*(blue and yellow)

9. *P*(red and red)

10. *P*(yellow and red)

Maria and Lucy are each shopping for a pet. Maria wants a puppy and Lucy wants a kitten. The table shows the number of puppies and kittens at the pet shop.

Puppies		Kittens	
Breed	**Number**	**Breed**	**Number**
Dalmatian	2	Burmese	2
Golden Retriever	3	Himalayan	3
Labrador Retriever	1	Persian	2
Poodle	4	Siamese	1

For Exercises 11–14, refer to the table. Find each probability.

11. *P*(Maria chooses a poodle and Lucy chooses a Persian)

12. *P*(Maria chooses a Dalmatian and Lucy chooses a Siamese)

13. *P*(Maria chooses a Labrador retriever and Lucy chooses a Burmese or Himalayan)

14. *P*(Maria chooses a retriever and Lucy chooses a Himalayan)

NAME ____________ DATE ________ PERIOD ____

Reteach

Probability of Dependent Events

If the outcome of one event affects the outcome of a second event, the events are called **dependent events**. The probability of both events occuring is the product of the probability of *A* and the probability of *B* after *A* occurs.

Example MARBLES **A bag contains 7 blue, 3 green, and 3 red marbles. If Agnes randomly draws two marbles from the bag, without replacing the first before drawing the second, what is the probability of drawing a green and then a blue marble?**

$P(\text{green}) = \frac{3}{13}$ 13 marbles, 3 are green

$P(\text{blue}) = \frac{7}{12}$ 12 marbles, 7 are blue

$P(\text{green, then blue}) = \frac{3}{13} \cdot \frac{7}{12} = \frac{21}{156} = \frac{7}{52}$

So, the probability that Agnes will draw a green, then a blue marble is $\frac{7}{52}$.

Exercises

1. Find the probability of randomly choosing a 2 and then randomly choosing an even number from a stack of 10 cards labeled 1 through 10. The first card is not replaced.

2. Two quarters, three dimes and seven pennies are in a coin purse. What is the probability of randomly choosing a penny and then a dime? The first coin is not replaced.

3. Lazlo's sock drawer contains 8 blue and 5 black socks. If he randomly pulls out two socks, what is the probability that he picks a blue sock then a black sock? The first sock is not replaced.

4. To pick sides for a basketball game the coach places 5 red cards and 5 blue cards in a hat. Each player takes a turn randomly choosing a card. What is the probability that the first and second cards chosen will be blue? The first card is not replaced.

Multi-Part Lesson 2 PART D

NAME ______________________ DATE __________ PERIOD ____

Skills Practice

Probability of Dependent Events

1. Four red and three blue marbles are in a bag. What is the probability of randomly choosing 2 blue marbles if the first marble is not replaced?

2. Two letters are randomly selected from the word PRIME. What is the probability that both letters selected are vowels? The first letter is not replaced.

For Exercises 3–6, use the information below.

A standard deck of playing cards contains 52 cards in four suits of 13 cards each. Two suits are red and two suits are black. Find each probability. Assume the first card is not replaced before the second card is drawn. Each card is chosen randomly.

3. *P*(black, red)

4. *P*(diamond, diamond)

5. *P*(jack, queen)

6. A card is chosen at random from a deck of 52 cards. It is not replaced and a second card is randomly chosen. What is the probability of getting a jack and then an eight?

7. Two cards are chosen at random from a standard deck of cards without replacement. What is the probability of getting 2 hearts?

8. A CD rack has 8 classical CDs, 5 pop CDs, and 3 rock CDs. One CD is randomly chosen and then a second CD is randomly chosen without replacement. What is the probability of choosing a rock CD then a classical CD?

9. A jar holds 15 red pencils and 10 blue pencils. What is the probability of randomly drawing one red pencil and then one blue from the jar? The first pencil is not replaced.

NAME ______________________ DATE ____________ PERIOD ______

Reteach

Make Predictions

A **survey** is a method of collecting information. The group being studied is the population. To save time and money, part of the group, called a **sample**, is surveyed.

A good sample is:

- selected at **random**, or without preference,
- representative of the population, and
- large enough to provide accurate data.

Examples **Every sixth student who walked into the school was asked how he or she got to school.**

School Transportation	
Method	**Students**
Walk	10
Ride Bike	10
Ride Bus	15
Get Ride	5

1 What is the probability that a student at the school rode a bike to school?

$$P(\text{ride bike}) = \frac{\text{number of students that rode a bike}}{\text{number of students surveyed}} = \frac{10}{40} \text{ or } \frac{1}{4}$$

So, $P(\text{ride bike}) = \frac{1}{4}$, 0.25, or 25%.

2 There are 360 students at the school. Predict how many bike to school.

Write an equivalent ratio. Let s = number of students who will ride a bike.

$$\frac{10}{40} = \frac{s}{360}$$

You can solve the equivalent ratio to find that of the 360 students, 90 will ride a bike to school.

Exercises

SCHOOL **Use the following information and the table shown. Every tenth student entering the school was asked which one of the four subjects was his or her favorite.**

Favorite Subject	
Subject	**Students**
Language Arts	10
Math	10
Science	15
Social Studies	5

1. Find the probability that any student attending school prefers science.

2. There are 400 students at the school. Predict how many students would prefer science.

NAME ______________________ DATE __________ PERIOD _____

Multi-Part Lesson 3 PART A

Skills Practice

Make Predictions

Determine whether each sample is a good sample. Explain.

1. 250 people at the beach in the summer are asked to name their favorite vacation spot.

2. Every fourth shopper at a grocery store is asked whether or not he or she owns a pet.

For Exercise 3–6, use the table and the following information. A survey of students' favorite sports was taken from a random sample of students in a school. The results are shown in the table.

Students' Favorite Sports	
Soccer	8
Baseball /Softball	3
Volleyball	5
Track & Field	4

3. What is the size of the sample?

4. What is the probability that a student will prefer soccer?

5. What is the probability that a student will prefer volleyball?

6. There are 550 students in the school. Predict how many students at the school prefer track and field.

For Exercises 7–10, use the table and the following information. A random sample of 40 flower shop customers was surveyed to find customers' favorite flowers. The table shows the results. The shop expects to sell 50 bunches of flowers on Sunday. How many bunches of each flower should the shop order?

Favorite Flower	
Type	**Shoppers**
Daisy	8
Gardenia	4
Mum	8
Rose	20

7. daisy

8. rose

9. mum

10. gardenia

NAME ______________________________ DATE ____________ PERIOD ______

Reteach

Problem-Solving Investigation: Act it Out

Example **Makayla thinks that by rolling a number cube, she can roll each number before Bobby can. Bobby wonders how many times he would have to roll the number cube to get each number at least once.**

Act it out to determine the simulation of rolling the number cube until each number is rolled at least once.

Understand You know that each person needs to roll the number cube until each number is rolled at least once.

Plan Use a number cube to complete the simulation. Bobby can roll the number cube and record the data in a table.

Solve Roll the number cube, and make a table of the results.

Outcome	Tally	Frequency
1	𝍸	5
2	|	1
3	𝍸 ||	7
4	|	1
5	𝍸	5
6	𝍸 |	6

After 25 rolls, each number, 1 through 6, was rolled at least one time. So, Bobby would need to roll the number cube about 25 times in order to roll each number at least once.

Check Check by doing several more trials to see if the results agree.

Exercises

1. **MONEY** Deshawn needs $0.65 for a vending machine. If he has several quarters, dimes, and nickels in his pocket, how many different ways can he make $0.65?

2. **ELECTIONS** Julia, Mason, Jackson, Ava, and Lashonda are running for class president and vice president. How many different possibilities are there for class president and vice president?

Skills Practice

Problem-Solving Investigation: Act it Out

Use the *act it out* strategy to solve Exercises 1–5.

1. **TOYS** A store is randomly giving away four different characters from a popular line of action figures. If Jerome receives one action figure each time he visits the store and he wants to get all four characters, how many times should he visit the store? Design an experiment and act it out to make a prediction.

2. **FLAGS** Mr. Miller's class is designing a flag for their school. They want the flag to have three horizontal stripes of the same width, one blue, one gold, and one white. How many different flags can they choose from?

3. **KITTENS** Alicia's cat had six kittens. Alicia wants to keep two of the kittens and sell the others. How many different ways can Alicia choose two kittens to keep?

4. **CASHIER** Paco recently started a new job as a cashier at a grocery store. Using quarters, dimes, and nickels, how many different ways can Paco give a customer $0.45 in change?

5. **TRACK** Four friends are competing against each other in a 50-yard dash. How many different ways can they finish the race?

Use any strategy to solve Exercises 6–9.

6. **FAMILY** The table shows the number of siblings for each student in a class. Make a table of the data. How many students in the class have more than two siblings?

Number of Siblings				
2	0	4	3	1
1	2	0	3	2
1	1	0	5	2
3	2	1	1	2

7. **NAMES** Molly wants to name her new puppy using two initials. If she picks two letters from the letters A, J, T, and B, how many different names are possible?

8. **SHOES** Lupita took a survey and found that about 4 out of 10 students wear sneakers to school more often than any other shoes. If there are 80 students in the sixth grade, how many sixth-graders wear sneakers to school?

9. **NUMBERS** Ian is thinking of two numbers between 1 and 20 whose product is 24. Find all the possible numbers.